D1233764

THE PHYSICAL SCIENCES SINCE ANTIQUITY

THE PHYSICAL SCIENCES SINCE ANTIQUITY

Edited by Rom Harré

ST. MARTIN'S PRESS
New York

Scholarly & Reference Division,
St. Martin's Press, Inc., 175 Fifth Avenue, New York, NY 10010
First published in the United States of America in 1986
Printed in Great Britain

Library of Congress Cataloging-in-Publication Data

A History of the physical sciences since antiquity.

Includes index.
1. Science — History. I. Harré, Rom.
Q125.H67 1986 509 86-12850
ISBN 0-312-38174-3

CONTENTS

PREFACE

This study was commissioned originally by another publisher as the first volume of a projected series of science textbooks. The series was to be built around the idea of different approaches to the understanding of matter. Unfortunately the project had to be abandoned for various reasons. Since this volume seemed worth salvaging we have revised and updated our contributions and present this joint work as, we hope, a useful background reader not only for those who in schools, colleges and universities are actually engaged in the study of matter, but also as a work of interest to the general reader.

Human understanding of the behaviour of matter, and our first inklings as to its nature, must have begun in the workshops, smithies, potteries and granaries of the ancient world. But the origins of a systematic science of matter seem to date from philosophical speculations amongst the Ancient Greeks. What records we have of these early thoughts on the nature of things suggest that it was the diversity of kinds of matter that first excited man's intellectual curiosity. Were iron, clay, water, smoke, wine, and so on, all manifestations of the same basic substance or was there a multiplicity of elementary materials? Some philospher-scientists among the Ancient Greeks argued for a monistic world of only one basic substance, others were prepared to accept the diversity of materials, with which we are presented in ordinary life, as an irreduceable fact of nature.

Along with the urge to reduce observable diversity to ultimate unity went the need to explain why some materials were soft, some hard, why some were medically useful and others inert, why some were inflammable and others hardened by fire, why some were liquids, some aeriform and some earthy. In this study we are looking at the many ways in which scientists from the remotest times have tried to explain the properties of matter. There has always been an intimate relation between philosophical speculation and scientific investigation, the one preparing the way for the other. Speculation often prompts the questions, while experimental investigation provides some of the answers.

The creative work of the critical disciplined imagination of theoretical scientists offers the experimentalist a repertoire of theories amongst which his practical knowledge enables him to choose, and with the help of which he directs his experimental researches. Our study shows how evolving ideas about the nature of matter became sufficiently clearly focused to allow for more and more careful and precise experimental tests.

The scientific study of the universe is not confined to elucidating the nature of matter. Cosmologists and astronomers have been interested in how the matter that makes up the concrete real world is distributed through the universe. Matter is structured in many ways; as galaxies, solar systems, the geological layers of the earth's crust, and so on. Cosmology is the science of matter on the largest scale, that of the universe itself. Only very recently has it turned out that cosmological theories about the origin of our universe can throw light on the nature of matter as it exists today.

Historians of science have long since given up merely chronicalling the discoveries that we, from our vantage point in the present, see as important or interesting. Great efforts are now made to try to write the history of science from the point of view of the times in which the events described in the lives and works of scientists were going on. The attempt at a greater realism has involved paying attention to the specific modes of thinking that animated researchers and led in the first place to an increasing interest in the philosophical element in science. But more recently it has become clear that in trying to understand the way scientists thought about their projects, and even how they directed their experiments, one has to see them as people very much of their own time and under the influence of social conditions of that time. The way society is organised, the economy, political changes and disputes, the wars and famines and so on, all in their different ways bear upon the projects that people think are worth pursuing, the problems they find intriguing and the kinds of solutions they agree are acceptable.

To tell a convincing and comprehensive story in the history of science requires that attention be paid both to intellectual and to social processes. Though the development of ideas is intimately interwoven with social and political matters, it is possible to some extent to treat them separately, in the later parts of our story at least. To this end the book is organised so that the history of speculation and experiment about the nature of matter is treated separately from the interplay between the sciences of matter and social forces. We have presumed that a reader needs to have a pretty good idea of how scientists of a period thought and some knowledge of the experiments that they did, before turning to the social influences that bore upon them in their time. So our sections on the social history of science follow and comment upon, illuminate and help to interpret the 'internal' history of the sciences of matter in a period.

R. Harré, Linacre College, Oxford

1 WHAT IS A SCIENCE OF MATTER?

Rom Harré

Scientific writings seem to be about two quite different kinds of matter of fact. There are reports on the results of experiments and observations, and these seem to be about things which each and every one of us could see, hear, taste, etc., with his own eyes, ears or tongue. Such matters of fact are said to be observable. There are many different kinds of observables. The colour of a brick, say, is an observable, and so is its length. We think of temperature as an observable property of matter, and so, too, electrical charge. We need no instruments to tell us the hue of a sample of red hot metal. Though we can feel warmth that feeling is not a good indication of temperature. Thermometers help us to make objective comparisons of degree of heat. But we could not even conceive of the idea of observing electrical charge without an instrument — the electroscope. Colour, length and temperature are all in some sense given in ordinary human experience of material things, electrical charge certainly is not. We detect it through its effects — a charged body changes the position of the gold leaves in the electroscope. That change has to be interpreted as the effect of change. It is not change itself. To make such an interpretation a theory is needed which postulates a causal link between an unobservable property of an electrically charged body and what happens when a suitable instrument is connected to it. This situation is quite typical of the physical sciences — temperature as a property of matter is the internal molecular energy of a substance, which causes a column of mercury to expand. Only the expansion is actually observed.

According to one account of the nature of science theoretical terms should be sharply separated from observational terms. The latter refer only to matters within a trained observer's experiences. Thus 'current', 'atom', 'electrical field', etc., are classified as theoretical terms, because though we may believe in the existence of these things we can never experience them as they are. 'Colour', 'speed', 'length' and the like are said to be observational terms, because we can experience them directly. This distinction is one of the founding ideas of a view of science called positivism.

There have been many instances in the history of science when many competent scientists were agreed as to the facts, as to what could be observed, but there were a multiplicity of theories available purporting to explain these observations. Was disease caused by a malfunction of the body which produced not only the symptoms but also the associated microbes, or were the microbes an invading army causing the malfunction? In the context of mid-nineteenth-century knowledge both theories were compatible with the observations and there was no more refined way of investigating the issue to come to some sort of decision as to which picture of the causes of disease was correct. Was the pattern of movements of planets, stars and moon and sun produced by a system of epicyclic motions of bodies wheeling around a stationary earth, or was the sun the physical centre of the planetary system, and the earth just another planet? There was no way that this issue could be settled in the sixteenth century and indeed there were no less than six geometrical schemes for the planetary orbits on offer by 1600!

Positivism is the theory of science that denies meaning to any but the observational terms. The multiplicity of theories available to explain some of the observable facts and regularities is interpreted according to the positivist account of science as a proliferation of rival ways of predicting new correlations among observables, rather than as rival attempts to depict a deeper reality. The positivist has little difficulty with the sixteenth-century puzzle of how there could be all those planetary theories, and is happy to accept Osiander's suggestion, in his preface to the great work Copernicus, that we choose the simplest theory, the one that makes our calculations easiest. There can be no question for a positivist of choosing that theory which is the best representation of reality.

This theory of science has appeared and reappeared since Ancient Greek thinkers first considered it. It has been associated with a certain, almost masochistic, attitude to the gaining of knowledge, a kind of deliberate amputation of one of our intellectual capacities, our ability to exercise our imaginations to pass beyond what we can observe to its hidden causes.

Positivism can be summed up in two important principles. One concerns meaning and one causality. Both are implicit in the above account. For a positivist the meaning of a statement about atoms, say, is reduced to what we would investigate to test whether the statement was true or false. So if a chemist talks about atoms, the meaning of his statements about atoms is to be sought in the relative and measurable weights and proportions of combining substances he would use to test his atomic hypothesis by examining its experimental implications. According to the positivist theory of science, statements about electric current are really only picturesque ways of talking about the pointer readings on the scales of appropriately designed instruments. The question of whether such entities as atoms or electrical currents exist or not is a misplaced question according to positivists, since even to pose the question supposes that the term 'atom' or the term 'current' has a meaning independent of the meanings of terms like 'weight', 'pointer-reading' and so on.

The second great positivist principle involves the reduction of causality from a productive relation between active agents and passive recipients of their action (acids act on bases) to a mere correlation of kinds of events. The correlations we observe induce mental habits in observers to expect the next event to be one of a kind already experienced in conjunction with what we take to be its cause. On this theory of causality we cannot suppose that there is really any causal influence or activity that produces the subsequent event. Taking aspirin is correlated with the disappearance of headache, but we cannot, on positivist principles, go on to say that aspirin actively reduces headache. That would be to mistake a mere psychological habit of expecting the headache to go after we take aspirin for a real relation in the world. The two great positivist principles are then the verification theory of meaning and the regularity theory of causality. For a positivist a theory of matter can be no more than a way of classifying the observable properties of material things.

Realists deny both these principles. For a realist the usefulness of a theory in explaining matters of fact already known and predicting new results suggests that the hypotheses of the theory and the picture of the inner workings of nature that it suggests should be taken very seriously as possibly true. So the reduction of the meaning of theoretical terms to the meanings of the terms used to describe ways of verifying theoretical statements must be denied. According to a realist 'atom' refers to a real constituent of matter, which — it just so happens — is not observable by a human being unaided by suitable equipment. How can a realist defend such a view? By choosing suitable examples a realist can show how theoretical terms have suggested, in certain cases, the existence of hidden things and processes that have actually turned up in further investigations. For instance, the atomistic theory of metals suggests that a metal object is a lattice work of electrically charged atoms (ions) which define the crystalline structure of the metal. When this idea was first introduced, perhaps by Berzelius nearly two centuries ago, there was no question of anyone observing the configuration of these ions. Indeed in the mid-nineteenth century there was a strong positivist movement in chemistry which sought to deny the meaningfulness of an atomic theory of matter altogether. But technical developments allowed observations of the ionic lattices to be made with great detail and precision. The lattices turned out to be pretty much like the atomic models that theoretically minded metallurgists had proposed to represent or picture the structure of matter that was hidden from ordinary observation by reason of its extreme fineness.

So, too, realists deny the regularity theory of causality. A realist does not accept that to say that citric acid dissolves calcium carbonate means no more than that there is a mere correlation between the presence of lemon juice and the dissolving of eggshells. Rather he believes that a very complex electro-chemical process has occurred with exchange of electrons, detachment of ions and so on. There is a causal process that links the conditions which initiate the causal chain with the effects that ensue. To prove the existence of such a process a realist has to use his theory of meaning, that the meaning of theoretical terms is not reduceable to that of observational terms, to suggest ways in which the existence and nature of the intermediate causal process productive of the effects could be studied. Sometimes these processes can be studied directly as in the oceanic currents that produce meteorological effects, sometimes indirectly as in the physical processes that occur when by means of an electric current a metal bound into an oxide is released. For a realist a theory of matter is an attempt at describing the true inner structures of material things, structures that explain the properties we observe things to have, such as their colour, chemical powers, weight and so on.

We shall see, in the historical sections, that scientists themselves have oscillated between the two main theories of science I have here described in interpreting their own theories about the nature of matter. For the most part science has been conducted in a realist framework, and efforts have been made to find the entities and processes mentioned in those theories which seem to have some degree of plausibility. But when difficulties arise, such as the appearance of many apparently equally plausible theories, or when it seems impossible to find a way of formulating a deep explanation, as has happened in some recent scientific developments, in particular the quantum theory of matter, then a positivist theory has often been favoured. Usually, though positivism is defended at the time with passion and even sometimes by the actual persecution of realists, there has almost always been a return to realism as new ideas and new techniques have become available. Difficulties again arise and a new swing to positivism begins, to be replaced by realism in its turn.

To understand what it is to have a theory of matter and how such theories are related to these views of science sketched above, we shall need to look more closely at how theories are constructed. The natural world of living and inanimate matter is not only extremely complex, but there are many features of interest it is easy to overlook. An example will help to illustrate the way scientists cope with these problems. There are many different ways one could study the air, and many different aspects of its behaviour on which one could concentrate. To make one aspect stand out from all others scientists usually employ an 'analytical model'. Robert Boyle discovered his famous law describing the way the pressure and volume of a sample of air were related by experimenting on gas trapped in the closed end of a bent tube. Why did Boyle choose to set up such an investigation? It seems that he was interested in whether the fact that there appeared to be no natural vacua, really empty spaces, at least in the region of the earth, was the effect of a physical process by which all incipient vacua were filled or a consequence of a sort of deep metaphysical principle, such as that expressed in the old adage 'Nature abhors a vacuum'! If the former was the case then creation of an artificial vacuum was just a technical problem, if the latter, it was impossible. So he conceived the idea that the air might act like a spring, and, as it were, spring back into empty spaces as they were being formed. He had therefore to compare the air to a spring. So he studied the changes in length of a trapped column of air under the influence of increasing weights. In a quite literal way he was investigating the spring of the air. He found that it did act just like a spring and that when the pressure was reduced below that of ordinary atmosphere the air expanded into whatever space was open to it. The behaviour of springs served as an analytical model for the behaviour of air. The first step in understanding the force of any theory is to try to pick out the tacit analytical model that has enabled the scientist to pick out of the complex web of reality some clear-cut phenomena to study. When Gilbert thought of the earth *as if it were* a spherical magnet (lodestone) he was able to identify what was really happening when ships used the magnetic compass to navigate.

But an analytical model only helps us to grasp the patterns and regularities in that part of the natural world we can observe. We have already run into the problems of how we are to think about things and processes which we cannot observe and yet which we believe to be responsible for the patterns we do see. Chemical reactions as we observe them, lead to changes in the colour, odour, crystalline shape and relative weights of substances. But why do these things happen the way they do? This is where another kind of model or analogue comes in. We cannot observe the hidden workings of nature, but we can imagine them. We have to try to imagine something that behaves in a similar way to the way we know that nature behaves. We know the way nature behaves from the use of analytical models to pick out observations of interest. To think of something that simulates the behaviour of real causal mechanisms we need to base our imaginings on something which we know does exist and whose workings we understand or think we understand. There is our source model. It is what we model our imagined causal mechanisms on. When a toymaker makes a toy car he models it on the real car, and the real car is the source model. It serves as a model for the design of the toy.

But when we have imagined the unobservable real natures of material things, we are still working only with an analogue. We suppose that nature is something like our picture, and then we see if our imagined causal mechanism would behave something like the way we know nature behaves. The representation of the hypothetical causes of natural patterns of behaviour is an explanatory model. It is only a model because it is just an analogue of the real mechanism whose behaviour we studied by using our analytical model. So a theory is a complex structure. Its heart is an explanatory model, which behaves like the real causal mechanisms of nature behave. This complex structure can be understood with an example or two. If we take the flow of a liquid as our source model for understanding electricity we can model our idea of electric current on it. But that idea or model is only an analogue of what is really going on in an electrical circuit. Its success depends on how far the laws of electricity, conceived as if they were the laws of a liquid flowing in a system of pipes, enable us to simulate the real mechanism's behaviour, which we have found out by using an analytical model, with the help of which we pick out observable relationships like Ohm's Law, linking and studying, say, its voltage and current (electrical pressure and electrical current, to speak in terms of the analogue). If we can get the law 'Voltage = Amperage × Current' by experiment, and if we can deduce 'Electrical pressure = electrical current × a constant' from our model then we have the analogy of behaviour that encourages us to think our theory plausible. Our model of the world behaves similarly to the way the world behaves.

This way of looking at scientific theories allows us to study the development of a family of theories. All the important connections between source model and explanatory model and between that model and the real processes are analogies. In the steady development of that family of theories we call 'the kinetic theory of gases', the idea of the gas molecule — the explanatory model whose behaviour had to simulate the behaviour of whatever gases are really composed — was modified several times in the light of new experimental discoveries, and these modifications were controlled by the use of a single source model for the whole series of theory modifications, namely that of a material particle obeying Newton's Laws of motion.

But this conception of theories also brings out the assumptions not usually made explicit in the way scientists write up their discoveries. In particular the analytical model that controls observation and the source model that controls theorising are usually not explicitly brought out.

As a general rule theories can best be understood as efforts to answer certain kinds of 'Why?' questions. To identify the task for theories of matter we should begin by detailing the kinds of questions which a theory of matter ought to answer.

The material things in our world are very diverse. There seem to be an enormous variety of stuffs out of which they are made and a number of phases or states in which these stuffs can be found. Water can be found as ice or in liquid form or as steam. Copper and clay and beef are all solid substances, yet they have very different properties and very different reactions. These simple examples suggest two of the most basic questions one could ask about matter. Why are there so many different kinds of stuffs? And why do they appear sometimes as liquids, sometimes as solids and sometimes as gases?

Related to these basic questions are some important subsidiary problems. Many things are composed of more than one substance. Bronze is made by fusing together copper and tin. Flesh seems somehow to be produced from grass and water. And sometimes not only are new substances formed by the bringing together of constituents, but those constituents actively engage one another, as acids combine with alkalis. What is the explanation of the activity of some substances and the passivity or inertness of others? If many common substances are composed of combinations (sometimes mixtures and sometimes compounds) of more basic materials, is there some kind of basic substance or substances out of which everything is finally made?

Much of our ways of answering such questions will depend on where we stand on the issues I have set out above, in particular whether we are realists or positivists by philosophical conviction. Realists will try to answer the questions by referring to unobserved structures and basic constituents out of which the observable materials are made. For them theory will emphasise explanatory models as possible representations of a hidden realm of things and structures whose behaviour will be held responsible for the observed distinctions and ways of behaving that are revealed in experiment and observation. Positivists are likely to try to answer the basic questions by attempting to formulate recipes for bringing about changes in the materials of the world and tend to treat such recipes as the ultimate laws of nature. References to unobservable atomic or sub-atomic structures could not be taken seriously by a positivist. At best they present us with dispensable aids to our weak and fallible intellects, at worst they delude us into thinking we know about strata of reality of whose very existence we must remain forever ignorant.

In what follows it will be clear that in practice most of those who have developed theories that try to answer the questions that the commonplace properties of material things force upon us think like realists. Despite setbacks and blind alleys they believed and still believe that they are penetrating beyond the surface appearances of things to their inner workings. So that when the differences between the solid, liquid and gaseous state of material substance is to be explained, part of that explanation consists in an analysis of the physics of the motion of its constituent molecules. How far are they bound into a structure that admits only of vibrations, and under what conditions are they energetic enough to break free into uncircumscribed motion?

But to be able to think in such a way a scientist needs to have at his disposal a model, or analogue, since he can by no means observe the dance of the molecules or the fluctuations of the magnetic field. Sometimes, as in the deepest theories of matter, a cluster of models is required to make the strange behaviour of the most deeply hidden constituents of matter intelligible to a limited human intellect. Part of what our history of theories of matter will relate is the history of the changing models or analogues in terms of which the facts that led to our questioning have from time to time been explained. With the help of such models our imaginations leap forward beyond where we can currently experiment into realms which we can know only by their effects on more mundane processes that we can observe either directly, like the melting of ice, or with the help of instruments, like the crystallisation of metals under different working conditions.

We begin with the very first attempts to provide systematic answers to the problems sketched above, the speculative science of ancient Greece.

Further Reading

R. Harré, *The Philosophies of Science* (Oxford University Press, Oxford, 1981)
J. Losee, *A Historical Introduction to the Philosophy of Science* (Oxford University Press, Oxford, 1972)
S. Toulmin, *The Philosophy of Science* (Hutchinson, London, 1953)

2 MATTER THEORY IN ANCIENT GREECE

Edward Hussey

A. The Rise of Greek Natural Science

1. Ancient Near Eastern Beginnings

Physical science began with the wish to know the future. The Babylonian priests and their royal masters believed that the movements of sun, moon and planets foretold what was to happen on earth. Over long stretches of time the priests observed these movements, accumulated records, noticed periodic patterns and tried to express the patterns by simple mathematical formulae.

There is nothing to suggest that the Babylonians tried to explain the regularities by any general theory. Their purposes were exclusively practical, while the belief that celestial bodies were gods, far above humanity, was not conducive to a search for explanations. Throughout the Ancient Near East, right across the arc from Mesopotamia to Egypt, general ideas about the structure of the observable universe, and about whatever might lie beyond it, seem to have been vague. None the less, speculations about origins appeared repeatedly: 'How did our observable world-order begin?' was a question often asked, and the answers carried implications for the stability or otherwise of the existing order. The answers were in terms of the activities of a creator-god or of several competing gods imposing order on an original chaos and coming to a political share-out of powers and responsibilities.

2. The Ionian Vision: a General Theory of the Universe

In the eighth century BC, the Greeks adopted an alphabetic script from the Phoenicians. This was not the only cultural borrowing they made from the Near East at this time. Over a period of some two or three hundred years, they learnt some of the cosmogonical speculation, some of the medical and commercial skills, and some of the mathematical and astronomical knowledge of the Near Eastern civilisations.

Greek science entered its original and creative period around 600 BC. No longer content merely to borrow ideas and knowledge from abroad, a few Greek thinkers, living in cities on the western coast of Asia Minor (the region of Greek settlement called 'Ionia'), began in the sixth century BC to do something quite new: to construct general theories of the universe. The fundamental innovation here was the setting of new standards of explanation which the theories had to meet. Traditional stories, Greek and foreign, about the origins of the observed world-order, had been criticised by the light of sober human reason and had failed to stand up. So the attempt was now made to provide accounts of the universe as a whole which were tailor-made to satisfy the demands of reason. That meant that, internally, they must be highly unified and coherent, with no arbitrary or *ad hoc* elements; externally, they had to account for the totality of the phenomena, in a natural way.

Such demanding standards of explanation were not fully and explicitly worked out all at once. Indeed, they did not become fully explicit until the work of Aristotle in the fourth century BC. But the vivifying presence of a new concept of explanation is already manifested from the sixth century BC onwards by the appearance of unified theories which have the unmistakeable 'feel' of a scientific construction about them.

Why was this all-important step taken just by these people at just this time? As already suggested, the possibility of free critical discussion of the traditional ideas is likely to have been essential. This freedom can hardly be unconnected with the political evolution of the Greek cities towards greater equality in the distribution of political rights and powers. Furthermore, the demand for more satisfactory types of explanation is characteristic of a politically and legally sophisticated society. In any case the Greek cities, unlike most Near Eastern societies, contained no great concentrations of wealth and power having a vested interest in the traditional beliefs. Free critical discussion was possible, and the results of reflection on these discussions could now, thanks to the spread of literacy, be written down and so preserved for further criticism and elaboration.

3. *The Development of Greek Natural Science*

Before we look at Greek theories of the material constitution of the physical world, it may be helpful to give a brief outline of the history of Greek natural science in general.

(a) The Sixth Century BC. The earliest attempts at the new style of cosmology were made by three citizens of Miletus in Ionia: Thales (active *c.* 575–545 BC), Anaximander and Anaximenes (probably slightly younger than Thales).

Only the broadest outlines of these cosmologies can be stated with any confidence (see section B 1 below). The attempt to expel arbitrariness and incoherence from explanations led to the conception of a unified and uniform universe, unbounded in space and time, essentially the same everywhere and consisting of a homogeneous stuff. (This is not to say that the concepts of space, time and matter, in terms of which we naturally phrase these theories, had yet been made explicit.) The observable world-order (*kosmos*) was not unique, but only one of infinitely many such systems, it was derived from the original stuff, and its structure and its phenomena were to be explained accordingly. Within this ambitious programme, bits of specialised astronomical and geographical knowledge make their appearance, though there is no good evidence of systematic study of astronomy or geography at this period.

(b) The Fifth Century BC. Beginning with Heraclitus of Ephesus (*c.* 500 BC), the original cosmological enterprise was repeatedly subjected to philosophical criticisms and reworked in various ways. This was part of the pattern for the next two hundred years, in which two distinct developments can be seen: (1) Natural science branched out from cosmology, and in the early fifth century comes the first appearance of scientific medicine, scientific astronomy and pure mathematics (all probably building on Near Eastern foundations), as well as proto-scientific speculations (with some attention to the empirical evidence) on biological and psychological questions; (2) at the same time, all this theorising was subjected to an ever-growing volume of ever more sophisticated philosophical criticism. While the original aims were not abandoned, the means thought appropriate to their realisation by different thinkers varied greatly. Pressing philosophical problems emerged: about the logical implications of existence and of spatial and temporal properties; about the justification of claims to knowledge; about the nature of science and of explanation themselves. In the course of the fifth and fourth century discussions, the concepts which now dominate formal and informal philosophising about science — such as those of space, time, matter, force, power, cause, explanation and science itself — began to take shape explicitly.

After Heraclitus, the principal physical theorists of the fifth century BC were Empedocles (*c.* 460 BC), Anaxagoras (500–428 BC) and above all the joint creators of the Atomistic theory, Leucippus (*c.* 440 BC) and Democritus (*c.* 460–380 BC). In the autonomous study of astronomy and of medicine, the greatest names are those of the astronomers Meton and Euctemon (active *c.* 440 BC), the medical theorist Alcmaeon of Croton (*c.* 450 BC) and the clinician Hippocrates of Cos (active *c.* 420 BC).

(c) The Fourth Century BC. In this century, the two tendencies noted above were carried even further: (1) Some particular sciences rose to new heights as autonomous fields of study — notably, mathematics (which now entered its first period of really great achievement), astronomy (both observational and theoretical) and biology; (2) but the old Ionian tradition of general cosmological and physical theorising was completely 'captured' by the schools of philosophy, which now began to exist as formal organisations. Of the philosophers who founded schools, one, Plato (428–347 BC), was hostile to the study of natural phenomena. The other great philosopher of the century, Aristotle (384–322 BC), was not, and he absorbed into his courses of instruction, within his own philosophical framework, large parts of the Ionian legacy. But his philosophical doctrines, while favourable to the empirical study of biology in particular, cramped theorising about physics, since they claimed to lay down *a priori* the form of all such theorising, and even some of its content.

The underlying reason why autonomous cosmological theorising came to an end was probably not the influence of philosophy, but an intrinsic loss of confidence and vitality (see section C below). It is clear, in any case, that the men who contributed most to particular sciences were either philosophers or under philosophical influence. The first great mathematician, Eudoxus (first half of the fourth century BC), who made brilliant contributions to theoretical astronomy as well, was an associate of Plato; while the outstanding biologists of the period were Aristotle himself and his pupil Theophrastus (*c.* 371–286 BC).

(d) After 300 BC. Theophrastus and his successor as head of the philosophical school founded by Aristotle, Strato of Lampsacus (died *c.* 268 BC), led the way in a promising movement towards more empirical observation and freedom from metaphysical dogmatism in physics. But this soon faded out. In astronomy, Aristarchus (*c.* 310–230 BC) put forward a heliocentric theory. This found little support, while mathematical astronomy based on the geocentric view culminated in the work of Ptolemy (active *c.* AD 130) — a great but isolated achievement. The same description fits the mathematical statics and hydrostatics of Archimedes (287–212 BC), whose purely geometrical work prepared the way for the invention of the calculus in the seventeenth century AD; and the geometrical optics of Euclid (*c.* 300 BC) and others. In pure mathematics, beside Euclid and Archimedes, there was also the work on conic sections by Apollonius (third century BC). But in general, from around 200 BC, innovative work in the physical sciences ceased, and this branch of study petered out as a subservient and comparatively

neglected appendage of philosophy (see also B 9 below).

B. Theories of the Material Constitution of the Physical World

The following survey takes Greek theories of the material constitution of the physical world in chronological order. The story is reasonably coherent, and certain recurrent themes give it continuity; but it is not possible to make it wholly self-explanatory. Lack of evidence obscures some important developments, and others can be fully understood only in the light of philosophical theories which cannot be explained in detail here.

1. The Milesians

The account of the universe given by the Milesians has been described in outline above (see section A 3(a)). There was a single basic constituent of the universe, which was homogeneous, infinitely extended and everlasting, and which was supposed to be both the basic stuff of the universe and the power-source for its transformations. For it was something alive and always naturally in motion (perhaps even intelligent: see below). The principal theoretical problem was to account for the formation and structure of world-systems (*kosmoi*) such as our own observable world-system. This was done by taking the basic stuff to be liquid or gaseous in its usual state; the world-systems could then be thought of as forming like bubbles in the vortices caused by its movements. The rotary movement could also be used to explain, within the world-system, the motions of the heavenly bodies, and, on the model of a centrifuge, the separation-out of different kinds of stuff (earth, water, air) from an originally undifferentiated one. But how could one undifferentiated stuff become several different kinds? The Milesians appealed to observed phenomena. Water seems to turn into air, via mist or steam; it solidifies into ice, and even might be thought, at Miletus, to be turning slowly into earth (the deposition of silt by the river Maeander was (and still is) changing the coastline noticeably). So it was plausible to suppose that, whatever the precise mechanism might be, all observable physical and constituents of the kosmos were different forms of a single stuff.

If such analogies might be taken to show that such transformations are possible, they do nothing to explain how they come about, nor why they occur just when and where and in the quantities they do. It is possible that some of the Milesians may have attempted to short-circuit these further questions by an appeal to an overall plan conceived and executed by the guiding intelligence of the universe (there are some indications that the basic stuff of the universe was, for them, not merely essentially alive but essentially intelligent and active). But obviously such a move could not be considered satisfactory for long; unless the plans and activities of any guiding intelligence had also been taken to be in principle rationally explicable, the original aim of explaining the universe rationally would have been abandoned. We do not know how Thales proposed to solve these problems, but in Anaximander we find the first appearance of the idea that at least the large-scale changes follow a 'law-like' periodic pattern in time. In fact Anaximander explicitly used the legal metaphor or model; he spoke of cosmic 'injustice' and of its being 'made good' by 'legal process' and 'penalty'. The construction is impressive in its dimly visible outlines, but the overriding requirement of explanatory unity seems to be in danger of being forgotten. Anaximander does not seem to have explained how his opposite forces, and his presiding principle of cosmic justice, could all have been derived from the one basic 'Unbounded' (as he named the basic stuff). Perhaps for this reason, the third Milesian, Anaximenes, reasserted the essential unity of all the physical components in a strong way: they were all just different forms assumed by air, more or less compressed. But how he then proposed to bring the transformations of air into a law-like pattern, is not known.

2. *Heraclitus*

Heraclitus, a philosopher of great originality, was aware of the problems encountered by Milesian speculation and tried to create a metaphysical framework within which they could be overcome. To reconcile the divergent tendencies of theoretical unification and diversification, he proposed a way of looking at the world in which a unity could reveal itself only through diversity and all diversity presupposed an underlying unity. What this amounted to in physical terms, when applied to the problem of cosmology, was that the universe possessed an organic wholeness (which he may have taken to be strictly analogous to the organic and self-organising wholeness of a human mind). The free, and yet in fact law-like, self-organisation of the physical universe was the work of an intelligent agent which he called 'fire', which manifested itself in a sequence of opposed physical states, not all of them manifestly fire-like. The underlying unity was expressed by the overall law-likeness of events; opposites were regularly transformed into opposites, according to cosmic 'justice'. Repeated regular oscillation between opposites is thus the characteristic pattern of physical change, which was worked out, probably, in some detail. The physically important opposites seem to have been the pairs hot–cold and moist–dry; the manifest constituents of the kosmos were determined as combinations of these — earth = dry + cold; water = moist + cold; air (i.e. cloud and mist) = moist + warm; (ordinary) fire = dry + warm.

The thought that the manifest forms involved in change are only expressed of an underlying unity led Heraclitus to introduce the first recognisable conservation principle: 'All things are an exchange for fire, and fire for all things, just as goods are for gold, and gold for goods.' Gold, the universal medium of exchange, is so only because it has a constant value in real terms; hence in transactions of exchange there is a conserved quantity of gold-value. Since 'fire' is the fundamental form both of physical stuff and of physical energy, there is a sense in which the conservation of mass, the conservation of energy and more generally the equivalence of matter and energy are foreshadowed in Heraclitus' physics. (Heraclitus did not, so far as we know, explicitly distinguish between mass-like and energy-like aspects of 'fire', and of course was far from having the concepts of mass and energy as they are now used.)

3. Parmenides and Empedocles: Elements, Compounds and Forces

In the first half of the fifth century BC, cosmological theorising took a fresh turn in the work of Parmenides and Empedocles. They recognised an absolute, irreducible plurality in the ultimate constituents of the physical world. In order to preserve some kind of overall unity none the less, they supposed the universe to consist only of a single kosmos, and they imposed upon this kosmos the unity of a living organism. In their vitalist theorising, they drew on the new attempts, in which they themselves were leading figures, to found a theoretical biology. (Parmenides is better known as a metaphysician who apparently questioned whether the observable world was real at all; but that does not affect his achievements as a physicist.)

The system of Empedocles is known in some detail. Here the classical four elements — Earth, Water, Air (or Ether) and Fire — appear as such for the first time. They cannot be transformed into one another. But they are capable of forming compounds. The concept of compounding (*krasis*) is an essential part of the theory. In a compound, the component elements lose their own charactistic properties, to a greater or lesser extent, and even their individual identities; they are present in the compound only in the attenuated sense that they can re-emerge from it. Each different proportion of the four ingredients produces a different compound, and Empedocles even specified the proportions in certain cases (e.g. bone is a compound in the proportions Fire 4, Earth 2, Ether 1, Water 1). All perceptible bodies are composed of elements and of compounds of elements. At least two of the elements, Fire and Water, are also physical agents. But to account for the formation and dissolution of compounds Empedocles found it necessary to introduce physical forces exterior to the elements. True to his vitalistic philosophy, he saw these processes as cases of 'love' and 'hate' among the elements. True to his explanatory ideals, since neither love nor hate can be taken as essential properties of elements (since elements are seen to exist both in compounds and uncompounded), he introduced, as two extra ultimate constituents of the kosmos, Love and Hate, much as a modern physicist might specify certain fields or types of interaction as fundamental, in addition to certain types of particle.

4. Anaxagoras: a Micro-structure of Potentialities

The second half of the fifth century BC saw two different attempts to bring the Ionian tradition of cosmology up to date. Both Anaxagoras and the early Atomists (Leucippus and Democritus: see section 5 below) had a concern, characteristic of their period, for the philosophical problems of knowledge, and were aware, after the impact of Parmenides' metaphysics and Zeno's paradoxes, of the philosophical problems lurking behind commonsense assumptions about what there is in the world. But they reached very different types of theory, diverging in a way analogous to the divergence between the 'bootstrap' and 'quark' theories of elementary particles (see Chapter 8). Both Anaxagoras and the Atomists tried to explain the observed facts about the compounding of ingredients, about nutrition and other phenomena in which chemical interactions play a part, by postulating a level of structure in material bodies which was not directly accessible to sense-perception. Their accounts of this 'micro-structure' differed radically.

In Anaxagoras' theory, there was no void. There were an unlimited number of different naturally occurring kinds of material stuffs. These all existed as potentialities within every volume of matter, however small. The manifest qualities of any given material object were supposed to be derived from the predominating potentialities within it; predominance, here, was conceived of in terms of qualities. It is, in effect, a sketch of a field theory of matter, the total field being the sum of infinitely many independent fields having positive strength everywhere. Anaxagoras evidently did not pretend to know, or try to investigate, how his fields evolved through time, except that he required that the total amount or strength of any kind of stuff should remain constant — a conservation principle again. It may even be that he thought there was nothing further that could be known by human intelligence about the matter. His pioneering effort was too obviously unable to proceed, given the mathematical and physical knowledge of the time, and was swiftly eclipsed without finding any continuation in the ancient world.

5. *Leucippus and Democritus: Early Atomism*

Early Atomism appears to be the invention of the elusive Leucippus, and to have been elaborated by Democritus, a prolific and wide-ranging writer. At the heart of Atomism is an axiom about what there is: atoms and void are all that exists. Void was a negative entity: unbounded, homogeneous, isotropic, unresisting to the passage of atoms, and unable to act upon them. Atoms were physically indivisible chunks, all alike except in size and shape. There were (probably) no theoretical limits placed on the variety of atomic sizes and shapes, though within any one world-system only a limited range of sizes and shapes would normally be found. Atoms were everlasting, impenetrable and in themselves unchanging. The only kind of change possible was the motion of atoms through the void, these motions being interrupted by collisions of atoms and consequent rebounds.

From this economical foundation, Atomism aimed to account for the whole of the observed phenomena, not only physical in the narrow sense but also biological and psychological; even a new theory of morals and politics was constructed by Democritus from Atomist principles. It was the first and still in some respects the most audacious attempt at thoroughgoing materialism.

It is natural to ask how far the early Atomists worked out the dynamics of atomic movement. A mathematical theory there almost certainly was not, but there are some traces of a qualitative theory, which may have required conservation of certain quantities without working out the mathematical consequences (an impossible task before the invention of calculus). Even this much is uncertain: incomplete and unsympathetic reports leave much obscure. It is possible that something like Newton's First Law of Motion was stated, and that concepts approximating to those of 'momentum' and 'kinetic enegy' were used.

In any event, the Atomists seem to have been more concerned to account for the formation and observed properties of our kosmos. Democritus devoted much effort to accounting for the sense-perceptible properties of the ordinary material furniture of our world on atomistic principles. In this enterprise he faced problems of various kinds

— problems belonging to philosophy and to psychology — as well as those that belong to the subject-matter of this chapter, which are the problems of constructing an Atomistic 'chemistry'. Democritus tried to bring the known facts about compounds (mostly derived from observations of cookery and metallurgy), the effects of heat and cold on bodies, and the interactions of bodies (including magnetism) into an Atomistic framework. He could use and adapt some earlier theorising — e.g. that of Empedocles (see section 3 above) — to whose four-element theory he gave an Atomistic underpinning by associating the four elements with four particularly common atomic shapes.

With the construction of the Atomistic theory, at the end of the fifth century BC, the great period of Greek 'natural philosophy' came to an end. Few really far-reaching ideas were to be produced in the remaining millennium of classical antiquity. From the point of view of a modern scientist, what was needed, around 400 BC, was a systematic effort to distinguish between competing theories by empirical investigations. The theory-making had fed for too long on a too slender diet of observations. There had indeed been a slow progressive realisation of the complexity of the phenomena to be explained, and this kind of progress was to be maintained for another hundred years at least. Even so, very little deliberate and systematic exploration of the natural world had been done, and not all that much more was ever to be done in Antiquity, except in a few isolated fields (see section C below). From 400 BC on, in spite of some exceptions, the general tendency is for no essentially new theories to be produced, for old ones to be reworked in the service of some philosophical school or other, and for discussion of rival theories to become mere partisan polemic between those schools. Little more than lip-service is paid to the need to account for the phenomena, and there is little evidence of a wish to investigate actively those phenomena. While the investigations of topics such as logic, theory of knowledge and philosophy of space and time became far more sophisticated, as organised philosophical schools came to dominate the intellectual scene, it is undeniable that the study of the natural world in itself fell into a slow decline.

6. Plato's Timaeus: *Geometrical Micro-structure*

Plato held that knowledge of the observable world had in itself no value, and that empirical investigation could not lead to any real truth. None the less, if only to round off his metaphysical theory, he gave an account of the general nature of the sense-perceptible world, which appears in his *Timaeus*. It is put forward as being only a 'plausible story' — which means, not that Plato was willing to revise it in the face of new evidence, but that he was uncertain how best to fit the observable world into his metaphysical scheme.

The cosmology of *Timaeus* was influential — because it was Plato's — in late Antiquity and in the Middle Ages. When it comes to the details of cosmology, far from trying to deduce everything from metaphysics, Plato in fact proceeds much like a theoretician of the late fifth century BC. True, he has a general philosophical position to preserve, but that does not by itself determine the overall shape of his physical theory. That theory presupposes the same questions as lay behind the work of Anaxagoras and Democritus, and above all the task of fitting Empedoclean element-theory into a more unified framework. As Anaxagoras and the Atomists had done, Plato solves the problem of providing a theory of the micro-structure of matter. Plato's theory draws on the ideas of certain thinkers who had seen themselves as followers of Pythagoras, notably Philolaus (active *c.* 450 BC). Philolaus, it seems, analysed physical bodies into a 'form-like' and a formless or 'matter-like' aspect. The form-like aspect of things was closely associated by him with the natural numbers and their properties, and it may be that he had already constructed a theory involving some kind of numerical or geometrical components in the micro-structure and seeking to explain observable facts in terms of mathematical properties of those components.

Plato, in any case, followed the lead of Philolaus, and (true to his saying that 'God is forever engaged in geometry') tried to make mathematical relationships play an essential part. For Plato too the fundamental duality is one distinguishing the form-like aspects of material things, which are geometrical shapes copied from the timeless world of pure mathematics, from their matter-like aspect, which in *Timaeus* is a single extended space-like matrix called 'the receptacle'. This receptacle becomes approximately 'formed' at different times and places by the impress of the geometrical shapes. The basic shapes are right-angled triangles, out of which regular polygons are built up, which in turn form the surfaces or regular solids. The combinations and transformations of the four traditional elements are explained in terms of the construction and dissolution of the polygons, the four elements corresponding to each of four of the five possible regular solids. The polygons, and the triangles of which they are composed, are thus not themselves material, but are permanent and re-identifiable possibilities for the construction of formed matter.

The whole theory is interestingly analogous in outline to the

General Theory of Relativity. The receptacle plays the part of the space-time manifold, and the geometrical shapes correspond to the metrical and other geometrical properties imposed on the manifold. The result of their combination is a 'formed' manifold representing the presence of physical bodies and forces. Plato's conviction that mathematical structures were at the heart of physics, which has been verified so strikingly in the present century, was based not at all on empirical investigations, but on his profound respect for the beauty, clarity, timelessness and generality of mathematical truth.

7. Aristotle: the Rejection of Micro-structure

The other great philosopher of the fourth century BC, Aristotle, was interested in every aspect of the natural world, convinced of the importance of physics and aware of the need for empirical investigation. Yet his theory of the nature of matter is less modern-looking than Plato's, though worked out in much greater detail and with much more attention to empirical evidence.

Most striking is what must look to us a retrograde step: Aristotle rejects any kind of micro-structure in material bodies. There seem to have been different and interacting reasons for this move. First, Aristotle had no difficulty, from his knowledge of the phenomena, in locating strong objections to each of the different types of micro-structure that had been proposed. Secondly, Aristotle thought that such theories were unnecessary, since all the phenomena could be accounted for more economically by the postulation of potentialities or 'powers'. The theory of Anaxagoras, as interpreted above (see section 4 above), had tended in this direction, but Aristotle, in working out a clear and explicit concept of potentiality, banished the notion that potentialities were present as ingredients in a mixture, in a state of suspended existence. For Aristotle, potentialities were just properties of the things that exhibited them, not further reducible. (There is nothing intrinsically anti-scientific in this notion of potentialities, or in a theory in which some potentialities are not further reducible. Nor is it correct to allege, as has often been done, that Aristotle did not try to work out the effects of potentialities in a quantitative way. The only thing wrong with Aristotle's theory was that it later turned out to be inadequate and was then defended, against all reason, by self-appointed partisans of Aristotle (not by Aristotle himself).)

The third motive which seems to have operated on Aristotle is less easily excused from the viewpoint of modern physics. Aristotle had a general doctrine of the nature of the sciences, which prescribed that all sciences were to be based, ultimately, on self-evident axioms about completely knowable objects. In the case of the natural sciences, this means sense-perceptible objects. Hence micro-structure (until the invention of the microscope, and then it is no longer micro-structure) is ruled out *a priori*. (This restrictive theory of the sciences is the really ominous side of Aristotle's physical theorising; just as much as Plato, he was determined that physics should be shown to be under the control of metaphysics, and autonomous only in a strictly limited way. The successful annexation of physical science by philosophy was to last, if we disregard isolated revolts, for something like two thousand years.)

Without micro-structure, Aristotle for his substantive theory of matter fell back on a familiar 'four element' scheme, reinterpreted in the light of his matter-form analysis. The four 'elements' — Earth, Water, Air, Fire — are the primary types of material body, and they get their characteristic form-like properties from the paired opposite qualities: hot, cold, wet and dry. By allowing change betwen these opposites, Aristotle makes a place for the change of one element into another. In one essay — which is preserved as the fourth book of the *Meteorologica* — the attempt is made to found an Aristotelian chemistry on these ideas.

But the four types of formed matter also have their 'natural places' and 'natural motions'. Heavy matter (Earth and Water) naturally moves downwards (that is towards the centre of the kosmos, that is towards the centre of the earth). Light matter (Air and Fire) naturally moves upwards towards the edge of the celestial region (in which a totally different kind of matter and a totally different physics is supposed to be found). This is clear evidence of an attempt by Aristotle to formulate mathematically expressed principles governing these motions. Since late Antiquity, it has been generally believed that he formulated a law of amazing ineptitude, according to which heavy bodies fall with velocities in direct proportion to their weight. As is well known, Aristotelians of the Middle Ages and the Renaissance tried to defend this law, which had already been severely criticised in late Antiquity and was decisively refuted by Galileo. But its attribution to Aristotle may well rest on a misunderstanding. It is quite possible that the general principle Aristotle intended to apply to actual bodies was a much more nearly correct one, supported by his own empirical research.

Aristotle also formulated other general physical principles. Among them are: (1) the first known statement of 'no action at a distance'; (2) a principle of 'continuity', stating that all changes are continuous; (3) conservation principles, requiring preservation of quantities of certain 'powers' in physical processes; (4) something remotely analogous to Newton's Second Law, relating the amount of a 'power' acting to the amount and rapidity of the resulting change. His physics was, then, a conceptually rich system of thought, well suited to be a starting-point for further investigations.

8. Theophrastus and Strato of Lampsacus: Empirical Physics of the Successors of Aristotle

The first two of Aristotle's successors in the leadership of his Peripatetic school took up the challenging task of improving Aristotelean physics. They criticised and rejected some of its more dubious features, and suggested, on the basis of empirical evidence, new forms of theory.

Theophrastus organised the collection of a large mass of observations relating to physics. A surviving essay of his, 'On Fire', uses observed phenomena to raise difficulties for Aristotelian theory but is not itself constructive. In his 'Physics', a longer work now lost, he advanced his own ideas. He rejected, it seems, the sharp division of the kosmos into two regions, the celestial and the sublunary (that is beneath the moon), and therefore the dual physical theory that went with the distinction; this was a promising start. The four elements remained the basis of the theory of matter, but one of them — Fire — was singled out as incorporating the active principle of heat, while the other three were taken to be essentially inactive. In the struggle between the activity of heat and the inertia of the rest of matter, the resulting transformations obeyed a law or laws of conservation. It is even possible (the evidence is unfortunately weak) that Theophrastus may have postulated a particulate micro-structure for bodies and the existence in them of scattered stretches of void.

Strato of Lampsacus continued Theophrastus' work, and enough is known to show that he extended it along the lines mentioned. In particular, it is certain that he postulated a micro-structure of material bodies akin to that of Atomism. In support of the existence of void, he appealed to deliberately contrived experiments. Both Theophrastus and Strato shed the clumsy Aristotelian theory of 'natural motions' retaining as fundamental only the downward tendency of heavy bodies. A nameless member of the school, who composed the surviving essay 'Mechanics' attributed to Aristotle, went so far (following hints in Aristotle himself) as to suggest analysing motions and forces into their rectilinear components. Here was one key to the development of a successful mathematical dynamics, which unfortunately could not be properly used until the development of calculus.

All this was clearly extremely promising. Unfortunately the promise of the early Peripatetic school was not fulfilled. Its interest in empirical physics faded out rapidly thereafter and had no enduring influence.

9. *Later developments*

Some few philosophers and others continued to be interested in physical theory after the third century BC, but for most of the rest of Antiquity, a period of some eight hundred years, there were few new ideas, little increase in knowledge and only sporadic activity.

The influential philosophical schools of the Epicureans and the Stoics treated physics as an adjunct to moral philosophy. The Epicureans put their faith in a reworked Atomism, a thorough exposition of which still survives in the great poem 'On The Nature Of Things' by the Latin poet Lucretius (first century BC). The Stoics reverted to a vitalistic four-element theory. Neither of these schools, it seems, did any empirical work on the subject, nor attempted to deduce precise laws of physical change. In the period of the Roman Empire, a revived Platonism came to dominate the philosophical scene among Christians and pagans alike. The neo-Platonists naturally revived the *Timaeus* theory of Plato, which being unspecific allowed of development. Among some neo-Platonists of the fourth and fifth centuries AD there flowered a new enthusiasm for physical theory, mixed up with attempts at astrology, alchemy and magic. New ideas were tried out about mechanics, the nature of physical interactions, and the role of mathematics in the physical sciences, some of them anticipations of modern developments. Unfortunately, the neo-Platonists, true to their Platonic allegiance, did not believe in the systematic empirical investigation which alone could have made their ideas truly fertile.

Outside the philosophical schools, there were isolated achievements in founding branches of physics which could be treated autonomously in a mathematical way: optics, astronomy, statics and hydrostatics. But no attempt seems to have been made to link any of these with general dynamics or matter-theory.

Alchemy deserves a special mention. As in its later history, it combined practical chemical and magical operations in varying proportions and was backed by a varying amount of theory, some at least superficially rational, some frankly occult. No attempt seems to have been made by anyone in Antiquity to unravel the genuine knowledge and the useful ideas from the rest and to found a science of chemistry.

C. Physical Science and Ancient Society

It is hardly possible to put forward well-based generalisations about the relationship between physical science and ancient society. In the first place, physical science, since it could not demonstrate practical relevance, was a highly marginal activity with no effects whatever on the economic or social structure. There is more hope of showing some influence of society on science. It is indeed very plausible, as has been suggested above (see section A 3), to connect the origins of physical theorising with the existence for the first time of a favourable conjunction of political, social and economic factors. Other connections have been suggested to account for the failure of ancient science to develop further than it did. The existence of slavery as a permanent social institution, the rise of Christianity and the loss of political freedom in the Hellenistic and Roman periods have sometimes been considered as contributing directly or indirectly to this failure. It is hardly possible to produce, with the evidence at our disposal and in the absence of agreement on the fundamental questions of sociology, a clear-cut case in favour of or against any one of these suggestions. But some points can usefully be made, using the history as given in section B.

If we focus the question on the arrested development and subsequent decline of physical theorising, it has already been suggested that the lack of systematic experimentation has some close connection with the problem. Why, then, did there never develop, in the ancient world, a tradition of physical science based on systematic experiments as well as theorising?

In the first place, it is clear that such a tradition was never envisaged by the founder-scientists, the Milesians. But, with the appearance of many competing theories, it might have seemed natural to appeal to deliberate experiment to decide among them. The fact that, instead, recourse was had to logical argument alone may reflect the difficulty of devising crucial experiments with the technology available; in addition, or alternatively, it may reflect the dominance of verbal argument in the scheme of higher education invented by the Sophists of the late fifth century BC and inherited by the philosophical schools as well as by those of oratory. The notion of higher education, since it was available almost exclusively to young men of the richer families, was naturally partly determined by conceptions of what it was proper for a Greek gentleman to know and do. However that may be, it was the philosophers who were in charge of what education in physics there was, from the fourth century BC onwards.

There was therefore an institutional bias against the experimental habit. But the institutional explanation seems to be insufficient, on at least two grounds. First, there was equally some institutional bias of the same kind in the early modern period, which was nevertheless eventually overcome.

Secondly, as mentioned above, it looks as though the bias was overcome already by 300 BC within the Aristotelian or Peripatetic school of philosophy, and converted into an institutional pressure towards experiment.

It is therefore the situation in the Peripatetic school under Aristotle's successors that first claims attention. Aristotle had set the pattern: active gathering of data in all fields of study was a part of his method. In biology, this extended to careful observations 'in the field' by Aristotle himself, and dissections of animals (as well as the collecting of a mass of countrymen's folklore and fishermen's tales). In physics, there is some reason to think that Aristotle had himself conducted simple experiments on falling bodies. Not only did his pupils keep this pattern, but they threw off some of the more restrictive of Aristotle's philosophical dogmas about natural science, sometimes using deliberate experiment to do so. With all this they created what could have been the beginning of an experimental tradition. But they did not leave it established: there was nothing to keep it in existence once their personal authority was gone.

All this at least shows that the non-appearance of such a tradition cannot simply be blamed on 'the philosophers' generally, nor on Plato in particular, whose influence, either inside the Peripatetic school or outside it, was small at this time. What is probably more to the point is to observe that the philosophies in the ascendant at this time were those of the Stoics and Epicureans, both of whom saw no point in treating physics as a field for autonomous investigation. It was moral philosophy that was demanded, for whatever reason, by the fashion of the time and perhaps by deeper causes having to do with the disappearance of the small traditional world of the Greek city-states and the rise of the Hellenistic kingdoms.

But this type of explanation in its turn cannot be considered wholly satisfactory. For in many fields besides general physical theory — ones which were not annexed by philosophy — we see the same kind of pattern. Around 300 BC and in the following century, there was excellent empirical work being done by astronomers (from which theoretical dividends were drawn by Aristarchus and Hipparchus), by mathematicians studying physical phenomena (optics by Euclid, statics and hydrostatics by Archimedes), by medical researchers in Alexandria (dissections and other investigations by Herophilus and Erasistratus). In all these cases, the work eventually faded out without establishing a tradition of experimental investigation, even though the institutions which had supported some of this work (the philosophical schools at Athens, the Museum at Alexandria) remained in being. And although there were still many gentlemen of leisure, with private incomes to support their studies, fewer and fewer seem to have studied the natural world for its own sake.

The problem remains. But it can be pointed out that, by comparison with, for example, Western Europe in the sixteenth and seventeenth century AD, the technological level of society was obviously far less of a help or a stimulus. In the third century BC, not only was technology less advanced in general, but instruments for the measurement of time were far cruder than they were to become, while the telescope and the microscope had not yet appeared to revolutionise human conceptions of the natural world. In general, expectations of further important scientific or technological discovery were very low or non-existent. Not only was there no widespread belief in the possibility of technical progress, but there was no large class of society who lived by technological skills, apart from very primitive and traditional ones. (Whether this has anything to do with the existence of slavery as an institution is dubious.)

Given the failure to establish an experimental tradition, the onset of stagnation, followed by decline, is most naturally explained as the result of the lack of new idea. By 300 BC all the plausible types of theory had been propounded, and later theorists did little more than shuffle them into slightly new combinations. The conviction then naturally spread that the only hope of understanding physics lay in starting from the right philosophical principles and consequently that physics was, and could be, no more than an appendage of philosophy.

Bibliography

Ancient Texts in English Translation

A large selection of important texts on all aspects of ancient physics is given in:
M.R. Cohen and I.E. Drabkin, *A Source Book of Greek Science* (Cambridge, Mass., 1958)
Translations of the more important surviving fragments of the early natural philosophers (down to Democritus), together with ancient testimonies about their thought, can most conveniently be found in:
G.S. Kirk, J.E. Raven and M. Schofield, *The Presocratic Philosophers* (Cambridge, 1983)
There are several English translations of the physical writings of Plato and Aristotle and of the poem of Lucretius, *De Rerum Natura (On The Nature Of Things)*. For example, Plato's *Timaeus* is available in the Penguin Classics series: *Plato: Timaeus and Critias*, translated by Desmond Lee (London, 1965). For Aristotle the most reliable translation is the 'Oxford' translation, originally published in twelve volumes at various dates under the general editorship of (Sir) W.D. Ross. This original edition is no longer in print, but a revised version of the whole translation has been published: J. Barnes (ed.), *The Complete Works of Aristotle: the Revised Oxford Translation* (2 vols., Princeton, 1984).

Lucretius is available in the Penguin Classics: Lucretius, *On The Nature Of The Universe*, translated by R.E. Latham (London, 1951).

Introductory Works

E. Hussey, *The Presocratics* (London, 1972)

G.E.R. Lloyd, *Early Greek Science: Thales to Artistotle* (London, 1970)

G.E.R. Lloyd, *Greek Science after Aristotle* (London, 1973)

S. Sambursky, *The Physical World of the Greeks* (London, 1956)

More Advanced Works

W.K.C. Guthrie, *A History of Greek Philosophy*, Vols. 1 and 2 (Cambridge, 1962 and 1965) (for very full coverage of physical theorists before Plato)

Kirk, Raven and Schofield, *The Presocratic Philosophers* (for up-to-date discussion of the texts)

G. Vlastos, *Plato's Universe* (Oxford, 1975) (for a recent account of Plato's *Timaeus*)

F. Solmsen, *Aristotle's System of the Physical World* (Ithaca, N.Y., 1960) (the most recent general treatment)

S. Sambursky, *The Physical World of Late Antiquity* (London, 1962) (a brilliant and pioneering exploration)

M. Jammer, *Concepts of Force* (New York, 1962) (chs. 1 to 3 deal with the ancient period)

Ancient Science and Society

The few books that can be recommended include:

J.P. Vernant, *Myth and Society in Ancient Greek* (London, 1980) (translated from the French)

J.P. Vernant, *The Origins of Greek Thought* (London, 1982) (translated from the French)

G.E.R. Lloyd, *Magic, Reason and Experience* (Cambridge, 1979)

3 THE TRANSITION FROM THE ANCIENT WORLD PICTURE

John Roche

Medieval Theories of Matter and Form

The Structure of Matter

Many of the theories of matter developed in Antiquity were further elaborated in the Middle Ages, following Islamic precedents, and given new directions. As we shall see the innovators of the scientific revolution of the seventeenth century drew heavily on medieval thought on matter and much of its content survived the introduction of strict methods of measurement into physics and chemistry.

Aristotle believed that the most basic form of transformation of matter was substantial change. That which persisted through and underlay all observable change of material stuffs was called by the scholastics 'primary matter'. For St Thomas Aquinas as for Plotinus (*c.* AD 207–70), this primary matter is pure potentiality and completely without form or actuality of any kind, unintelligible in itself and known only indirectly. Even the individuality and quantity of a particular corporeal body, such as a piece of rock, was a consequence of the actualisation of primary matter in that body, as such, and not an innate property of primary matter itself. This extreme conception of primary matter as pure potency was strongly disputed throughout the thirteenth and fourteenth centuries. Although the arguments were conducted within an Aristotelian framework of analysis, they were much influenced by the Augustinian tradition. According to the latter matter was in itself the seat of active potencies or 'seminal reasons' in virtue of which it possessed substantial forms in a germinal state. Versions of this theory, somewhat attenuated, were supported by John Pecham (d. 1292), Duns Scotus (1266-1308) and William of Ockham (*c.* 1284-1349). All of this discussion arose because a careful analysis of Aristotle's teaching revealed obscurities, inconsistencies and unsolved problems. For example, how could matter be both the principle of individuation — that is, what made a thing that particular thing, as Aristotle seemed to suggest — and yet survive a change which resulted in a new individual?

Aristotle's theories of the structure of matter at more perceptible levels of organisation were as widely followed as his notions about ultimate matter. Gross matter in the sublunar world was composed of various combinations of the Empedoclean elements earth, water, air and fire. Each of these in turn was the result of a substantial union involving different pairs of the four quantities — hot, cold, moist, dry — with primary matter. Change in the sublunar world was ultimately brought about by the passage of the Sun through the ecliptic, which could even transform one element into another, a doctrine which encouraged alchemy.

Aristotle's doctrine of the four elements was discussed extensively in the Middle Ages. A true mixture of the elements was a true substance with its proper substantial form inhering in primary matter, and not a mere juxtaposition. The problem arose of how could the mixture be truly said to be composed of the elements, or in what manner were the elements present in the mixture? Aristotle had suggested that they were 'virtually' present but it was by no means clear what this meant. According to the Islamic scholar Avicenna (980-1037) the forms of the elements do persist in the new substance but in such a weakened state that they are capable of fusing and dispose the subject to acquire the new substantial form. This won acceptance in certain circles but philosophers mostly rejected it because it conflicted with the established Aristotelian principle that only one elemental form could inhere in matter at one time. Averroes of Cordova (1126-98) argued that both the substantial forms and the qualities of the elements are weakened in the process of mixing and are fused into the form of the mixture. This conflicted even more strongly with Aristotelian doctrine in that subtantial forms do not admit of 'intension' or 'remission'.

Averroes' theory had various supporters in the Latin West who modified it in various ways. Roger Bacon regarded the weakened elemental forms as a phase of gradual transition from potency to act.

Thomas Aquinas was more strictly Aristotelian in that he maintained that in a true mixture the element forms are indeed lost, and their qualities preserved only in the sense that during the process of mixing the qualities acting on each other generate a resultant 'middle' quality, which in turn disposes the matter to receive the new form of the mixture. In Theodoric of Freiberg's (*c.* 1250–1311) version of Averroes' theory even primary matter is predisposed in varying degrees to each of the four elements and the quasi-spiritual character of these dispositions allow simultaneous co-existence of the elemental qualities in a new homogeneous substance. In this analysis of medieval discussions of combinations of the Aristotelian elements it is perhaps possible to glimpse a characteristic which the seventeenth century was to find so distasteful: a rigorous and detailed logic is applied to an obscurely defined and empirically vacuous terminology.

Greek commentators on Aristotle had developed the theory that each substance had its own natural quantitative minima: if the substance were divided below that size it would cease to exist as that substance. This did not imply a corpuscularian theory of bodies. Indeed, for Aristotle, each proper substance was a continuum. According to Averroes, minima sometimes actually exist. They were the first things to come into existence when a body is generated and the last things to be lost when it perishes. When reacting substances are mixed, according to Averroist doctrine, the chemical interaction takes place between the minima themselves which fuse, according to the doctrine of mixtures, to become temporarily the minima of the new structure. During the thirteenth and fourteenth centuries the Averroist doctrine of natural minima was modified in various ways but retained its essential character. One new problem introduced was whether there are qualitative minima as well — e.g. whether a given colour can exist below a certain intensity.

The various condemnations by the Church in the thirteenth century of details of Aristotle's theories gave medieval thinkers a freedom to criticise and a flexibility of analysis which was to prove very fruitful. Aristotle's denial of atoms, the void, infinity and plural worlds was subjected to extensive criticism. Philosophers argued that God could create a body moving in empty space. Nicholas of Autrecourt (d. after 1350), in the context of discussions of the Eucharist, adopted a fully Epicurean Atomism. Though some writers, including Nicholas of Autrecourt, maintained that vacua probably exist, most followed Aristotle and denied that a vacuum could actually exist in Nature. Various experiments were discussed to illustrate the physical impossibility of a void. According to Roger Bacon's (*c.* 1249-92) famous formulation, 'in a vacuum nature does not exist'.

A very different theory of the ultimate structure of matter — which is now known as light metaphysics and which was derived ultimately from neo-Platonism through various Islamic and Jewish sources — was given prominence and further developed by Robert Grosseteste (*c.* 1170-1253) of Oxford. According to Grosseteste during the Creation the self-diffusion of light through unformed primary matter had generated all bodies. Furthermore, light is the first effect and cause of its subsequent developments. This theory was to have a profound influence on magnetism, on the theory of gravity and even, perhaps, on modern field theory.

The relation between matter and its quantity received considerable development in the thirteenth and fourteenth centuries. Quantity as denoting the internal disposition of the parts of a body (perhaps to some extent comparable with the modern concept of structure) was distinguished from quantity as shape (to be studied in geometry) and as spatial extension. The development of a concept of 'quantity of matter' was motivated by both theology and dynamical analysis. It was an ancient Christian principle that matter, created by God, cannot be generated or annihilated by any creature. This theological principle of the conservation of matter received further support from the belief in the resurrection of the body, since the matter of the latter must continue to be available in the universe, in order for the original body to be reconstituted at the Last Judgement.

The need for the concept 'quantity of matter' was felt strongly in the fourteenth century. It was postulated as that which persisted in any condensation or rarefaction, and as that which explains the different resistance offered to motion by two bodies of the same size but of different materials. The concepts of density and rarity were refined from common experiences, such as compaction and expansion, and were thought of as directly intelligible and quantitative. Though 'quantity of matter' and 'density' were therefore informally quantified in thought by medieval analysts, procedures of measurement were never specified. Newton was to make use of just such informal quantitative concepts in his early definition of quantity of matter or mass.

Medieval Dynamics and the 'Latitude of Forms'

Another important medieval concept, developed as a result of the critical analysis of Aristotle's physics, was that of 'impetus'. According to Aristotle's theory a projectile was in unnatural or 'violent' motion and, as with all motion, required an immediately contiguous agency to sustain the motion. Aristotle had rather vaguely suggested that the original propellant imparts to the surrounding air a power to move the projectile which is transmitted through the air with the mobile. This was felt to be highly implausible in late Antiquity: beating the air violently ought, on this theory, to cause a stone to move, but it did not.

A more popular answer in the fourteenth century, which derived from late Antiquity, was that of impetus theory. According to this theory the projector imparts a dynamic quality to the projectile, the 'impetus', which internally sustains the motion of the body. According to Jean Buridan at Paris (*c.* 1300-58) impetus was proportional to both the quantity of matter of the body and its velocity. Furthermore it would continue to act steadily unless dissipated by resistance. The heavenly spheres, having no resistance, continue to move for ever under the action of the impetus originally given them by God.

Medieval discussions of weight and motion — and indeed of changes in degree of any quality — frequently employed a concept of quantity which requires careful understanding. Today we think in terms of numerical measurement by means of a universal standard. The schoolmen did not usually think of their quantities as measured in numbers by means of standards. Such measurements were for astronomers and tradesmen. They generally represented their quantities by lines, and thought of them in similar parts, comparable with each other as lines were in the geometry of Eudoxus and Euclid. That is, any physical quantity was thought of as having a certain proportion to any other similar quantity, as lines had to other lines. Interest was centred on the magnitude of the proportion or ratio, rather than on the absolute values of each of its terms. This proportion, it was supposed, could be measured in principle, just like that between lines, by simple mutual comparison or by adding equals to each term of the proportion until both agreed. The impracticality of doing this with degrees of sickness and virtue did not deter medieval thinkers from analysing such concepts quantitatively, using the geometry of proportions as their model. Furthermore, many properties could, in principle, be treated validly in this manner, such as force, velocity, hotness, acceleration and density. It is now possible to recognise that the fundamental weakness in these investigations was the absence of practical and tested measuring definitions for the quantities discussed.

With various forms of notional quantification as a basis medieval scholars, especially in the fourteenth century, developed an elaborate theory of matter based on the idea of the 'latitude of forms'. This was a mathematical theory of the degrees of a quality and first appeared in Antiquity in medical and pharmacological writings. The 'latitude' of human health and sickness was quantified, for example, using the four Aristotelian qualities, hot, cold, moist, dry, each of which was given four degrees. By the thirteenth century, this mode of analysis had spread to philosophy and theology. Its application to the science of motion and other physical qualities was first developed at Oxford early in the fourteenth century. A distinction was made between a quality uniformly distributed over a body in space or time, such as uniform motion, and qualities non-uniformly or 'difformly' distributed. A special class of the latter was a uniformly difform distribution in which the quality increased or decreased at a constant rate.

The technical language of the theory of latitude of forms gave us such expressions — still in use — as 'uniform velocity' and 'uniform acceleration', and 'extensive' and 'intensive' quantities. Applying these conceptions to kinematics William Heytsbury (*c.* 1313-72) argued that a uniformly accelerated body corresponds to its mean degree of velocity — that is it travels the same distance in a given time as it would if it moved at its mean speed all the time (the Merton 'mean-speed' rule). Nicole Oresme (d. 1382) developed a far more elaborate theory of the latitude of forms, applied to motion, explicitly using diagrams to represent the degree of intensity of quality by a vertical line and its extension in space or time by a horizontal line. The area of his figures was given physical meaning by Oresme as the 'quantity of a quality' and contained important anticipations of Galilean and Cartesian mechanics.

Esoteric Sources of Matter Theory

Alchemy

The theories of matter so far discussed have been severely rationalist. However, this presents a one-sided view of medieval thought on matter which was also strongly influenced by the esoteric and the occult. This is well exemplified in medieval alchemy which, as a body of theory and praxis, was chiefly derived from Islam. Alchemy appears to have arisen in late Antiquity in Alexandrian Egypt in the union of the practice of Egyptian metal workers with the theories of matter of Alexandrian Gnostics and neo-Platonists. Alchemy, and eventually chemistry, came to occupy the middle ground between the physics of gross matter and the physics of ultimate matter. The possibility of transmuting one substance into another was based on Aristotle's doctrine of the transformation of the elements, although the primary matter underlying the change was thought of by alchemists as a substance in itself. The fundamental object of alchemical practice was to transmute the base metals copper, iron, lead into the noble metals gold and silver. In the course of their efforts to achieve this, ancient and medieval alchemists, as well as incorporating traditional metallurgy, dye-making, cosmetic and other processes, established a large body of chemical processes and apparatus and discovered many new chemical substances. All of these processes and substances were explained within a framework which saw the visible universe as abounding in natural symbols of deeper powers and which included among the forces of nature sympathetic action, action at a distance, celestial influence, the power of numbers and magical powers in natural objects. It was an animistic view of the universe which was keenly conscious of cosmic sympathy and antipathy and the parallelism between the macrocosm, the universe, and the microcosm, man.

Under the influence of natural magic the concept developed — particularly in Islamic alchemy — of a single refining agent which when 'projected' onto a base metal was held to be capable of transmuting it into gold. This was sometimes called the 'philosopher's stone'. According to alchemical theory the dry and moist exhalations — which, condensing together, Aristotle believed produced the metals — in fact first produce sulphur and mercury, and metals are formed by a subsequent combination of these subtances. The most perfect natural harmony produces gold. A defective union produces the base metals. The philosopher's stone was held to foster the perfect growth of sulphur and mercury into gold. Sulphur was identified with the male, active and spiritual principle, mercury with the female, passive and corporeal principle. The Work of the alchemist (the Magisterium) was seen as closely parallel to human sexual intercourse, conception and growth of the foetus. Astrology was indispensable to the success of the Work because of the supposed natural affinity between the planets and various metals. Because of the belief in a parallel between man (as microcosm) and the universe (as macrocosm) alchemy was also considered important for medicine. The philosopher's stone was held also to cure imperfections of the human body.

Medieval alchemy in the West took over the writings and practices of Islamic alchemy and added little that was new to the body of theory. However the chemical substances, apparatus and processes handled by the alchemists were substantially extended to include improved methods of distillation, new methods of chemical analysis (to recognise the presence of gold), and an increased use of the balance of chemistry.

Magic

Another unlikely source of theories about matter, also involving a strongly experimental approach, was magic. This flourished in late Antiquity deriving its beliefs and practices from all of the Near Eastern and European cultures. Magic involved much experimental skill requiring years of practice with substances of all sorts and including much chemical expertise. These practices were embedded in a somewhat incoherent body of doctrines involving an underlying belief in a sympathy or affinity among all things animate or inanimate. The correct invocation of the names of gods or potent humans was held to animate certain substances with magic powers. The natural magical properties of certain stones, herbs, and innumerable other substances, and their divinatory powers, could be evoked by the use of certain secret preparations, words, symbols, diagrams, numbers and actions. Magic (and alchemy) was chiefly concerned with the control of natural forces and human destiny, rather than with developing a detailed understanding of natural processes. It was primarily an art, therefore, rather than a science.

Natural magic was sharply distinguished from evil or daemonic magic and sorcery, for example by William of Auvergne (*c.* 1180-1249) and Roger Bacon. The discovery of occult virtues was a principal concern of many medieval experimenters who included the real and supposed powers of the magnet as part of natural magic. Physical and occult causes were regarded as equally capable of producing physical phenomena, a mode of thought from which seventeenth-century science had to struggle to emancipate itself. Nevertheless there are strong links between the systematic investigative approach to the control of nature developed in the seventeenth century and the earlier unsystematic but experimental quest for occult powers, natural marvels, and ingenious artefacts as a means of exercising a human control over nature.

Magic also flourished even among the learned during the sixteenth century, however the latter was careful to draw a distinction between natural magic and sorcery. Cornelius Agrippa (*c.* 1486-1534), one of the strongest supporters of magic, in 1530 wrote that 'Natural magic . . . [was] nothing else, but the chief power of all the natural sciences; which therefore they call the top and perfection of Natural Philosophy . . . which by the assistance of natural forces and faculties . . . performs those things that are above Humane Reason'. John Dee (1527-1608), an Elizabethan scientist and magus, was equally enthusiastic about what he termed the art of Archemastry: '. . . because it proceedeth by experiences, and searcheth forth the causes of conclusions by Experiences . . . it is named of some Scientia Experimentalis'.

During the sixteenth century the forces which magic undertook to investigate in order to achieve control over nature, continued to be, as in the Middle Ages, sympathies and antipathies; the natural signs revealing inner powers; the medical and other virtues of stones, gems, organs, herbs and plants and liquids and powders; magnetic attractions; optical illusions and mechanical and other natural or supposed wonders of all sorts. The magical and the naturally marvellous were not distinguished. However, the later sixteenth century is noticeable also for a shift to a more critical and investigative approach towards claimed natural marvels and away from a concern with the occult and the superstitious. This is most noticeable in the work of *Natural Magic* of 1558 and 1589 of Giambattista Della Porta (1555-1615) who, alongside a miscellany of natural curiosities and artefacts, tested experimentally and falsified the claim, for example, that garlic weakens the lodestone. William Gilbert's *De magnete* (1600), the classical study of the properties of magnets, can indeed be seen to some extent as a final purification of this tradition: the subjection to systematic experimental investigation and inductive explanation of part of the traditional content of natural magic.

Natural Philosophy in the Renaissance

Nature as a Mathematical Text

Alongside the uncritical interest in the occult and the fantastic in the sixteenth century, other more rational lines of thought were developed, sometimes in the same mind, which culminated in the new scientific movement near the end of that century. One such development was the growing interest in Platonism in Italy both inside and outside the universities. The Platonic view that nature is at bottom mathematical and that it is constructed according to definite mathematical proportions, became quite widely accepted, as did the consequence that a scientific understanding of nature required a mathematical training. This contrasted strongly with the established Aristotelian view which assigned to mathematics a secondary role, and to quality and essence a primary role, in physical theory. Archimedes became the ideal of the platonically inspired inquirers into nature, with his rigorous Euclidean style of reasoning and his mathematical manipulation of simplified, idealised physical quantities. Even the most mathematical of the Aristotelian corpus of texts on physics, the *Mechanical Problems*, was far more intuitive and discursive in style than Archimedes. Despite much propaganda in favour of the postulational and mathematical approach of Platonism and Archimedes, however, little was actually achieved in this direction until the writings of Simon Stevin (1548-1620) and others. If this were the only development in scentific method in the sixteenth century, however, it is doubtful whether it would have achieved much more than an extension of those subjects amenable to the Archimedean method, which was not experimental.

The Rational Arts and Crafts

A very different approach to nature, however, developed in northern Italy from the fifteenth century, largely outside the universities and spread to the rest of Europe. This resulted from philosophical reflection on the improvements in the arts and crafts which were so striking a feature of the age and stimulated expectations of similar improvements in theoretical disciplines. Many of the traditional technologies were improved by rationalisation and empirical trials, and many of the useful and pleasing arts were advanced by the systematic application of mathematics and measurement. The mathematical theory of perspective was applied to painting and technical drawing. Geometry and proportion were systematically used in architecture, sculpture and military fortification. Astronomical theory and instrumentation were applied to navigation principally, at first, by the Portuguese. Surveying, cartography and gunnery employed more sophisticated mathematics and instruments. Machine-making, horology and mathematical instrument-making also formed part of what came to be known as the rational, mathematical arts.

Many scholars in northern Italy involved themselves in the practice of art or engineering, and many artists, a class originally of unlettered craftsmen, cultivated scholarly interests. The best known are perhaps Brunelleschi (1377-1444), Leone Alberti (1404-72), Leonardo da Vinci (1452-1519) and Michaelangelo (1475-1564). As a result there developed a most profound analysis of the relation of the mind and manual activity to the control of nature by art, and also of the relationship of art to nature. Art here was understood in its broadest sense as the production of any human artefact. This scholarly experience in manipulating matter, the discovery that design in the mind had to be refined by trials with refractory matter, generated a far more empirical attitude towards the understanding of nature among these philosophical artist-engineers. In the course of the sixteenth century, in certain circles, nature began to be seen increasingly as a work of divine craftsmanship. The world, indeed, began to be seen as a mechanism and, just as a piece of clockwork is best understood by dismantling and reconstructing it, the notion developed that nature could best be explored and understood by its manual investigation. In the Platonic, and even in the Artistotelian tradition, a reasonably close attention to nature was thought to be a sufficient empirical ground for theorising about physics. The new approach to nature through art was to carry Western science far beyond its achievements in Antiquity. Instead of attempting to understand the world by reconstructing it from rationally postulated and empirically remote ultimate principles, it sought to discover by manual investigation immediate structures in nature and then to work 'downwards' and 'upwards'. This is still a central commitment in the Western investigative approach to nature. The joining together of instrumentation, exact measurements and systematic experimental exploration, with postulation, induction and mathematics, with varying emphases, were to determine the chief styles of scientific method in the following century.

The Composition of Matter

The two philosophical movements just discussed, the Platonic-Archimedean and the rational arts, prepared the scientific mind in the sixteenth century but did not yet add much to the theory of matter. The various classical, Islamic and medieval theories of matter were also extensively discussed during that century. However, little that was fundamentally new was added to earlier theories. It seems worthwhile, nevertheless, to summarise briefly these discussions since in some cases new concepts were introduced. Furthermore, they form a background and sometimes a starting point for the advances made in the following and in later centuries.

Sixteenth-century discussions of motion and mechanics are noticeably different in style from their fourteenth-century antecedents. They often display vigorous interaction between the established scholastic tradition, the reviving Platonic and Archimedean approach and the eminently practical style of the rational artists. Many medieval writings on the science of weights and on motion were printed, and concepts such as that of impetus, the latitude of forms and positional gravity were widely discussed during the century.

In the sixteenth century discussions of the combinations of the elements and of natural minima were not advanced much beyond similar discussions in the Middle Ages. Many of the Arabic and Latin medieval alchemical texts were published in the course of the century, reflecting a continued interest in and acceptance of old doctrines. There was also a substantial body of literature attacking the claims of the alchemists. There did appear, however, in the early sixteenth century an offshoot of medical alchemy which included an original and influential theory of animate and inanimate matter. This was the outcome of the efforts to reform medicine of the bizarre and prolific Philippus von Hohenheim, or Paracelsus (1473-1541). He insisted that the main purpose of chemistry was not with the transmutation of metals but with the preparation and purification of chemical substances for use as drugs in medicine. The theoretical framework in which Paracelsus embedded these useful notions was hermetical, cabalistic and magical. The usual microcosm–macrocosm parallelism in terms of a correspondence between Spirit–Soul–Body, with Logos–Nous–Universe was extended by Paracelsus to metals and eventually to all substances. He conceived every substance to be built up of three principles: mercury, sulphur and salt — mercury being the active and spiritual, sulphur the mediatory and salt the passive corporeal principle. These principles replace both the Aristotelian four elements and the four humours of Hippocratic medical theory. Man's health depends on the equilibrium between these three principles, which can be restored, when lost, by drugs prepared chemically according to the same three principles. This theory formed the basis of both paracelsian iatrochemistry, or drug chemistry and of his conviction that chemistry is called upon to reveal the hidden workings of nature.

Paracelsus was faced with the same problem which had appeared in the Aristotelian theory of the elements, as to how his principles persisted in a chemical compound. He maintained that the threefold material principles — mercury, sulphur and salt — persist unchanged in the compound but are unified and organised by a fourth, more spiritual principle, the Archeus. This, though conceived in a Stoic manner as a more refined sort of material, meant that for Paracelsus all matter was somehow animate. The occult and esoteric character of much of Paracelsus' doctrines won him few adherents among philosophers and considerable hostility from the medical profession. Nevertheless his iatrochemistry was considerably influential among physicians and chemists and initiated a tradition of research which made important contributions to *materia medica* and chemical praxis.

What is usually called the scientific revolution gathered considerable momentum towards the end of the sixteenth century and may be regarded as a sudden flowering of many processes of maturation that had gone on throughout the century. It may also be regarded as part of an active response to the growing scepticism of the age towards all received systems of thought both philosophical and religious — a scepticism often accompanied by admonitions to consult the book of nature rather than the texts of Antiquity.

Bibliography

A.C. Crombie, *Augustine to Galileo*, 2 vols. (Harmondsworth, 1969)
E. Grant (ed.), *A Source Book in Medieval Science* (Cambridge, Mass., 1974)
M. Boas Hall, *The Scientific Renaissance 1450-1630* (London, 1962)
D.C. Lindberg (ed.), *Science in the Middle Ages* (Chicago, London, 1978)
C.W. Shumaker, *The Occult Sciences in the Renaissance* (Berkeley, 1972)
W.P.D. Wightman, *Science in a Renaissance Society* (London, 1972)

4 THEORIES OF MATTER IN THE SEVENTEENTH CENTURY

John Roche

The Development of a Mechanical Conception of Matter

The Properties of Solids, Liquids and Gases

The year 1586 is perhaps the most appropriate to mark the beginning of the new physics of matter. This was the year in which Simon Stevin (1548-1620), an engineer and mathematical philosopher, published the *Elements of the Art of Weighing* in Flemish. It was soon translated into Latin and French. In this work Stevin applied the methods and axiomatic style of Archimedes to the theory of weights and to liquids at rest, and surpassed both Antiquity and the Middle Ages in logical rigour and in the number of problems solved. By applying the scholastic principle of the impossibility of spontaneous or perpetual motion to a 'wreath' of spheres slung around a wedge, Stevin was able to deduce the law of positional gravity — that is the component of weight acting along a plane — in a more satisfactory way. In hydrostatics he showed that the thrust of a fluid on the bottom of a container was directly dependent on the depth of the fluid, on the area and on the specific gravity of the liquid. He also showed that the thrust on a portion of the side was the same as if it were horizontal. He solved correctly the problem of determining the thrust of a fluid on an inclined bottom. He did not, however, formulate a technical concept of pressure as the thrust per unit area in a fluid. In the theory of machines he made use of the principle that a balance is in equilibrium when the weights are in inverse ratio with the perpendicular distances to the vertical through the fulcrum, a concept familiar in medieval statics. This is still quite far from the concept of the turning action or the 'moment' of a single force about a fulcrum, as measured by taking the product of the numerical magnitude of the force and the numerical measure of its perpendicular distance to that fulcrum. Stevin also recognised that in systems of ideal pulleys and other machines in near equilibrium the distance covered by 'power' and 'resistance' are in inverse ratio to these forces. Again this is still far from an explicit formulation of the quantified concept of the work done by a single force. Such concepts only gradually matured in the course of the seventeenth and eighteenth century.

It is to be noted that Stevin's statical theory did not require any exploratory or confirmatory experiments. Common knowledge of behaviour of balances in equilibrium, or of fluids at rest, was a sufficient control over theorising and indeed lent a self-evident or *a priori* character to his axioms. However, even the idealised and simplified Archimedean analysis of nature could not get much further without an appeal to experiment to decide between postulates, or even to investigate the very beginnings of a problem and discover new concepts. In the same work Stevin describes how he and a friend dropped leaden balls to test Aristotle's law of falling bodies. He also carried out many other experiments. The combination in Stevin of engineer and natural philosopher, of a postulational and mathematical science of nature with experimental exploration and testing makes of him a practising example of both new styles of scientific investigation, which were to be articulated by others in the seventeenth century.

In the course of the seventeenth century the science of statics of solid bodies received considerable development following the Archimedean rational style. G. Roberval (1607–75) in a publication of 1636, the *Traité de Méchanique*, solved the problem of the equilibrium of three forces by applying the theory of the bent lever to it. In 1638 Galileo (1564–1642) published his *Two New Sciences* in which, in a dialogue form, he directed the attention of science for the first time to the problem of internal strains and breaking stresses in loaded beams, a topic previously confined to artisans and quite central to an adequate theory of matter. From these beginnings there gradually evolved, in the course of the next two centuries, concepts of internal tension and strain in a stressed material, defined in a measurable way, and the definition of internal stress as a force per unit area.

The developed concept of pressure applied both to the atmosphere and to the pressure in liquids arose out of attempts to create and 'measure' a vacuum, and also to explain it. Many writers in the sixteenth century had defended both the intelligibility and the actual physical possibility of a vacuum. Galileo in the *Two New Sciences* went further and discussed an experiment suggested by G. Baliani (1582–1666) for measuring the force of the supposed 'horror vacui'. He considered a cylindrical vessel filled with water, closed by a wooden piston, and upturned. By measuring the weight by traction needed to separate the piston from the water he proposed to put a definite measure on the 'horror vacui'. It is doubtful whether Galileo ever carried out this experiment.

In 1963 E. Torricelli (1608–47), a pupil of Galileo, had a similar experiment carried out under his instructions which involved upturning a long glass tube completely filled with mercury over a dish also containing mercury. The mercury slipped down the tube, always maintaining a fixed height above the mercury in the dish. This famous experiment led to extensive discussions and disputes over the nature of the space above the mercury. Some claimed that it was a true vacuum, but the Peripatetics and Descartes maintained that it contained a subtle matter which was responsible for supporting the mercury. Torricelli himself was in no doubt that the mercury column was balanced by the external atmosphere and that we live submerged at the bottom of an ocean of elementary air possessing weight.

The question was only to be settled by a series of elaborate experiments, conducted by Blaise Pascal (1623–62), which followed a rigorous logic of induction and testing. What finally decided the matter was the famous experiment on the Puy de Dôme, near Clermont, in 1648, where it was shown that the level of mercury in the upturned tube when on the mountain top was lower than when at its base. The best way to explain this was to assume a change in weight of the atmosphere with height. Pascal, reflecting upon these experiments, developed a single science of air and fluid hydrostatics. By means of his new principle, that fluid pressure is transmitted equally in every direction, he was able to reduce to a single explanation such varied phenomena as floating bodies and the hydraulic press.

Matter in Motion

It was in the understanding of the kinematic and dynamical properties of matter, rather than in any theory of its internal structure, that science in the seventeenth century achieved such notable advances.

It was Galileo who was largely responsible for the necessary modifications to the Archimedean method which allowed it to be applied to the study of moving bodies. To achieve this he had to join postulation and mathematical idealisation with controlled experiment, numerical measurement and deductive testing. Furthermore, he had to elaborate a body of new explanatory mechanical concepts. In his seminal investigation of falling bodies Galileo abstracted from the cause, gravity, and concentrated instead on the kinematics of the motion, that is on the acceleration, velocity, distance and time of motion of a freely falling body. He also abstracted from friction and air resistance in arriving at his principles. By rolling spheres of different materials down inclined planes and by constructing pendulums of the same length but of different materials, Galileo corroborated his postulate that all bodies of whatever material would fall at the same speed in a vacuum. By timing the path of a rolling sphere on an inclined plane Galileo confirmed his deduction that velocity of fall under gravity is proportional to the square of the time and indirectly confirmed his postulate that the acceleration is uniform in time. These results appeared in a developed form in the *Two New Sciences* (1638) but had been arrived at only with considerable intellectual effort over a long period. It is to misunderstand the whole methodological tradition to which he belonged, and was then reconstructing, to suppose that Galileo's laws of free fall were the outcome of induction from experiment. For Galileo, in the Platonic tradition, experiment had a necessary but secondary role to that of reason which was more creative and provided deeper and truer insights than the senses alone. Furthermore, he believed he had arrived at principles which carried both mathematical and physical certainty and that further experiment in this subject was unnecessary.

Perhaps the most important consequence of Galileo's discovery was that it sharpened awareness that bodies in (ideal) free fall were continually accelerating and that it no longer made sense to speak of the 'velocity' of a falling body as if it had a single value. Far from ignoring the cause of the acceleration Galileo recognised that it was the weight of the body, or for a body rolling down an inclined plane, it was the component of its weight along the plane. As a result of his mediations on the implications of this Galileo demonstrated that a body rolling onto a horizontal plane from an inclined plane will, in the absence of resistance, continue with uniform motion for ever, since there would be neither causes of acceleration nor of slowing down. 'Furthermore . . . any velocity once imparted to a moving body will be rigidly maintained as long as the external causes of acceleration or retardation are removed.' Uniform motion for Galileo thus became a natural state which contrasted strongly with the Aristotelian theory of projectile motion and also with the theory of impetus. In Galileo's day the most recent contemporary exponent of classical ideas about motion had been Kepler. He had argued, following the neo-Platonic devaluation of matter, that '. . . opposition to motion is a characteristic of matter; it is stronger, the greater the quantity of matter in a given volume'. This property of resistance, not only to the initiation of motion, but to acquired motion, Kepler called 'inertia'. Galileo had also earlier believed that the impetus acquired by a body naturally wastes away. In the *Two New Sciences*, however, Galileo is fully confident that a moving body has no innate tendency to resist its *acquired* velocity and only resists acceleration. It was his sharp distinction between the properties of velocity and acceleration that made this insight possible. Galileo's law came to be known as the 'inertial' law only because, in the hands of Newton, 'inertia' was more narrowly defined as a power to resist changes in velocity.

Since uniform horizontal motion was now conceived as natural there was no conflict within a body between uniform horizontal motion and accelerated free fall. He could therefore superimpose the two motions without mutual interference or destruction.

> Imagine any particle projected along a horizontal plane without friction . . . and if this plane is limited . . . then the moving particle will, on passing over the edge . . . acquire in addition to its previous uniform and perpetual motion a downward propensity due to its own weight; so that the resulting motion . . . is compounded of one which is uniform and horizontal and of another which is vertical and naturally accelerated.

This principle immediately led Galileo and others to the deduction of a parabolic trajectory for a projectile. It swept aside an ancient Peripatetic doctrine that contrary motions (that is natural and violent) could not inhere simultaneously in the same subject, and it was the starting point for innumerable advances in kinematics and in dynamics. It was, for example, Galileo's insight that weight causes uniform acceleration and not just motion in an undefined sense, and certainly not uniform motion, that eventually led to the general recognition that forces cause uniform acceleration and to Newton's second law relating force and acceleration in an exact way. It should be pointed out, incidentally that 'natural uniform motion' was, for Galileo, in a circle centred on the Earth. In this way the circular motions of the planets were also natural and required no sustaining intelligence or other causes. The conflicting contrasts between Galileo's and Kepler's views on motion are quite remarkable.

In statics Galileo made an important advance by conceiving the turning action of a weight as a distinct property of that weight in given circumstances, a property for which he coined the term 'moment' or 'inclination'. Galileo attributed various properties to bodies at rest and in motion. The suppression of a static balancing force liberated the 'impeto' or tendency of a weight to move downwards. He also drew attention to a power in bodies already in motion which is quite close to the concept — later to be more fully developed — of momentum: 'absolutely equal weights, moving with unequal velocities have unequal powers, virtues, momenti, the most powerful is the one that is most rapid, and that in the ratio of its own velocity to the velocity with which the other weight is endowed': Again he says that 'momento signifies that virtue, that action, that effective power by which the mover moves and the mobile resists'.

It was René Descartes (1596–1650) who first clearly expounded the concept of 'quantity of motion', defined it as proportional to the amount of matter and its velocity and stated that '. . . there is a certain quantity of motion in the whole of created matter, which never increases or decreases'; and so when one body sets another in motion, 'it loses as much of its motion as it imparts to the other'. Quantity of motion was not a dynamic power for Descartes but a state which was simply handed on from one body to another during impact. All physical forces, for Descartes, were reduced to a succession of such impacts, a concept he possibly obtained from Isaac Beeckman (1588–1637). In the course of the seventeenth century the Galilean concept of 'moment' or 'momentum' as a virtue or power was gradually identified with the Cartesian 'quantity of motion'.

Descartes took an important step in the recognition of 'work' as a distinct process, and towards its absolute quantifiication, by representing it as a rectangular area. One side represented the force, the other the distance moved. He also anticipated the concept of power or rate of working, representing it by a three-sided figure. Although Descartes paid considerable attention to the laws of impact between bodies, the rules he formulated were mostly false. Descartes' algebraic geometry, a culmination of efforts by Viète and Fermat, prepared the techniques which were to be used in the application of algebra to the motion of bodies and fluids.

Descartes must also be credited with one of the earliest deductions of a functional physical law. This was the law of refraction of light which he deduced from his theory of matter and light. Galileo had, of course, achieved a similar deduction in his law relating velocity and time. Such reasoning was to become increasingly a characteristic of postulational physics. Descartes also formulated the principle that 'rest' is of the same nature as 'motion' and a principle of the relativity of motion according to which a body can be said to be in motion or not according to the system of references chosen. Christiaan Huygens (1629–95) was to use such a principle in solving correctly the problem of the impact of two bodies. Newton, however, distinguished sharply between absolute and relative motion.

Huygen's was the first to employ an explicitly algebraic language for the static moment of a force and for momentum. He formulated correctly the law that 'when bodies collide the same quantity of motion is always conserved in the same direction'. He also explicitly stated that in elastic impact 'the sum of the products of the magnitudes of each hard body, multiplied by the square of the velocities is always the same before and after collision'. This product, which Huygens did not name, was to be later regarded as the measure of a new power in moving bodies, the '*vis viva*'. The novelty and the conceptual importance of this new mathematical idiom deserves more attention from historians of science than it usually gets. Here the language of proportion is abandoned and absolute measures are implied which gave a more sharply quantitative significance to the concepts employed.

Gottfried Leibniz (1646–1716) maintained that the true motive force or power in a moving body was not its quantity of motion as defined by Descartes but must be 'measured by the violent effect it produces' — that is by the height to which it can raise the moving body. He correctly pointed out that the height which two moving bodies can reach, if directed upwards, is not proportional to their respective quantities of motion but to the products of their quantities of matter and their velocities squared. Leibniz called this property of a moving body the living force (*vis viva*) and contrasted it with 'dead' force — that is a force in equilibrium. He maintained that it was *vis viva* that was always conserved and not the quantity of motion. He was not aware that the direction of motion had to be taken into account to get the correct law of conservation of momentum. However, the disputes with Huygens and others had the positive effect of giving *vis viva* a status of equal importance with quantity of motion as a property of moving bodies. Leibniz introduced yet another property or rather process in a moving body which he called 'action' and defined as the product of the *vis viva* and the time, or as the product of quantity of motion and displacement. This concept is, of course, extensively employed in modern physics.

It is worthwhile pausing to reflect on the various mechanical properties of matter to which the seventeenth century had been led in its analysis of bodies in motion. The more prominent ones included inertia, quantity of matter, quantity of motion, *vis viva*, action, velocity, weight and acceleration. Of these properties, inertia, momentum, *vis viva* and action are very much like scholastic hidden properties, except that they are sharply quantified and more strongly suggested by experiments. Their specification by algebraic products further concealed the fact that they were unobservable and not manifest properties. Yet Leibniz seems to have been quite unaware of the hidden character of the dispositional properties or power in his own theory when he attacked Newton's conception of gravity as 'occult' in his famous correspondence with Samuel Clarke.

The term 'force', with a range of meanings specified by the context of utterance, was of course of ancient usage. In its mechanical meaning it referred to a manifest property or rather a transitive process active in various static and dynamic circumstances, such as the support of heavy loads, traction and falling weights. In every case, except that of falling bodies, the agency or body exerting the force could be clearly identified, even with magnetic attraction. Physical conditions observable in most forcing agents also manifested the presence of a force, such as the spinning threads of a taut rope, or the bending of a supporting beam.

However, it was precisely the acceleration in free fall whose cause could not be observed that was to lead ultimately to an understanding that force causes acceleration. In the Aristotelian system also the cause of accelerated fall was occult, innate to a body, and not an external agency. It was surely conceptions of force of this sort that led Descartes to banish force as a distinctive property from his cosmology. Leibniz's employment of the term force to designate both a cause of motion and strain and an occult power in a moving body, together with the metaphyisical framework of his discussions of force, also helped to obscure the term considerably.

In the following century Jean D'Alembert (1717–83) was to call force 'obscure and metaphysical' and attempt to banish it from mechanics, in one of many positivistic reactions to recondite physical concepts. Nevertheless, the traditional empirical concept of force survived and was given increasingly precise specification by Isaac Newton (1642–1727) and Leonhard Euler (1707–83) and William Thomson (1824–1907).

It was his recognition of the extension of terrestrial gravity to the Moon that led Newton to realize that falling bodies are attracted, or at least are acted upon, by an external force which accelerates them. This also led him to the idea that common forces also generate acceleration and to his famous causal law that forces are proportional to the rate of change of momentum that they generate. This was not an empirically and inductively established law for Newton, but the outcome of a complex postulational union of Galileo's mechanics with his own theory of gravity. It indicates the sophistication with which Newton employed the Archimedean and Galilean methodology.

The concept of quantity of matter is, as we have explained, a medieval concept and much use was made of it throughout the seventeenth century. The habit of thinking comparatively, say between two bodies of the same material differently compressed, or of different volumes, allowed the relative amounts or quantities of matter to be easily understood in such cases. In Descartes' system 'quantity of matter' would simply mean the real volume occupied by his underlying homogeneous extension — substance. A similar meaning could be attached to the concept in an atomic theory. The concept of quantity of matter in the seventeenth century was not therefore as obscure as it was in the scholastic tradition or as it was subsequently to become. Its special importance in dynamics was frequently recognised in that century. G. Baliani (1582-1666) distinguished gravity as the 'agent' and the material body as the 'patient' and supposed them to be proportional to each other; a body does not fall more rapidly when it is heavier because the greater gravity has to set in motion a larger quantity of matter.

E. Mariotte (1620–84) realised that it is not the accidental property weight but the quantity of matter in a body that determines the quantity of motion of that body. By quantity of matter he understood the 'volume with a certain solidity or condensation of parts of its matter . . .'. This is very reminiscent of Newton's early definition of quantity of matter or mass as 'proportional to the density and to the bulk conjointly'. Newton was here, of course, employing the informally quantified medieval conception of density. In a later improved version Newton gave an exact operational definition of the mass of a body as proportional to the inertia or resistance it offered to being set in motion by a given force. The measure of a dispositional property, inertia, was therefore simultaneously taken to measure something deeper and more occult, mass, which was treated as the common seat of both inertia and gravitational action and, eventually, of substance of that body. A comparison of weights was taken by Newton to be a comparison of masses. Newton introduced considerable confusion to mechanics, which still persists — by identifying quantity of matter — with its connotations of volume, and mass, which is quite unrelated to volume and which is specified dynamically.

Modern disputes about whether or not Newton's three laws of motion are mere definitions or truly empirical laws often fail to take several historical, physical and logical considerations into account. The law of inertia in physics from its inception has had the status of an inductively verified law, that is, it has been verified in every circumstance in which it has been tested and makes no absolute claim to hold necessarily in the future. It is used to predict the presence or absence of force but without necessity and certainly not to define or characterise a force. Force in physics is an empirical process that can usually be witnessed and which, among other things, causes acceleration. A force is not thought of as the acceleration itself.

With regard to Newton's second law it should be pointed out that Newton himself thought of it as a causal law and not as a definition.

Furthermore, no concrete problem in mechanics can be solved by definitions alone. Empirically authenticated laws are always needed, together with initial conditions. Even the simplest problem in dynamics requires the application of Newton's law as a functional relationship between observationally distinct and independent physical quantities. The process of discovery through which Newton arrived at his second law is distinct, logically, from its subsequent empirical justification, which is unimpeachable. In fact, it is this empirical law that makes possible and underwrites the choice of 'mass x acceleration' as an indirect measure of force in addition to the traditional direct measures with balances and springs. To say that force 'is' mass x acceleration is to ignore this and to commit the category error of equating a mathematical manipulation with a physical force, which is an observable transitive process going on in the world.

Theories of the Ultimate Constituents of Matter in the Seventeenth Century

Atomism

In the search for a more intelligible structure of matter to replace the increasingly descredited Aristotelian cosmology, many thinkers in the late sixteenth and early seventeenth centuries adopted some modification of classical Greek atomism, or of the scholastic conception of 'minima' or 'smallest parts'.

One of the earliest exponents of a new systematic theory of nature involving the existence of real minima and the void was Giordano Bruno (1548–1600) who expounded his views most fully in *De triplici minimo* (1591). Bruno engaged in a radical and metaphysically based rejection of the whole Aristotelian cosmology. He believed in an infinity and plurality of worlds: any point in infinite space may be taken as the centre, without any one being really so. There is no privileged place. The empty space that contains bodies is indifferent to motion. There can be no motion without a void. Although obscure in manner and militantly heterodox in the content of his works, for which he was burned in 1600, Bruno paved the way for that 'geometrisation' of space in the seventeenth century which gradually led to its being thought of as passive, inert and neutral. This was a very different conception from the polarised space of Aristotle.

Bruno spent the years 1583-85 at Oxford and London, during which time he published five works. His ideas were widely discussed in England

and there is some evidence that Thomas Harriot (1560–1621), who knew of Bruno's ideas, found certain of his notions stimulating, although Harriot was not given to obscure metaphysics. Harriot was one of the first mathematical philosophers in the seventeenth century to employ a thoroughly classical Atomism in his scientific explanations. Harriot's Atomism was closely allied to the mathematical problems of the infinite and of infinite division of the finite. However he also employed the concept of physical atoms, which he held to be eternal, indestructible, continuous and interspersed with void. Physical qualities result from the magnitude, shape and motion of these atoms. Dense bodies differed from rare bodies in having fewer void spaces interspersed between the atoms. Harriot explained simultaneous refraction and reflection by a transparent surface using his atomic theory. Refraction was a series of zig-zag internal reflections. Harriot did not publish his theory, however it appears to be well-known in the circle of Henry Percy (1564–1632), the ninth Earl of Northumberland, and may have contributed to later Atomic doctrines in England.

From about 1603–12 Francis Bacon (1561-1626) showed a considerable interest in an Atomic view of nature. Bacon's atoms are endowed with 'matter, form, dimension, motion, place, resistance' as well as 'appetite and emanation'. For Bacon, all atoms are identical seeds which act upon each other at a distance. Bacon was concerned to purify the Atomic doctrine and rid it of its traditional association with atheism. He argued that a 'Divine Marshall' was more necessary to explain the order brought out of an army of infinite small portions or seeds, than would be needed to explain the generation of the world from the five elements. Democritus' doctrine of the eternity of matter did not conflict with the Christian teaching that God created the universe out of a pre-existing matter or chaos.

Sometime before 1620 Bacon discarded a specifically Atomistic view of matter but retained the idea of bodies having internal structure, expressing this in the phrase 'latent process and configuration'. This was presumably as a consequence of his elaboration of an inductive philosophy according to which theorising is closely controlled by experiment and testing. The Greek atomic theory was, after all, theorising in the old style, producing conjectural explanation without precise deduction and without strong prompting by experimentation.

Pierre Gassendi (1592–1655), an outspoken adversary of scholasticism and a priest, was the most systematic exponent, in the seventeenth century, of Greek Atomism. Gassendi, as Harriot did earlier, adopted Epicurus' principle: *De nihilo, nihil* (nothing emerges from nothing).

It follows that there exists an imperishable prime matter common to all bodies. This prime matter is divided into atoms which are indivisible, identical in essence, and are extended, full, impenetrable and interspersed with void. Atoms were endowed by Gassendi with weight, that is with a power of self-motion. Gassendi, like Bacon before him, was concerned to reconcile Atomism with Christianity, and he held that atoms were created by God, who had the power to annihilate them, but no natural force could do so. Gassendi modified Classical Atomism to the extent that he attributed a hidden principle to the Earth, analogous to a soul, which activated chains of atoms which he believed to be responsible for gravity.

Atomistic theories of various sorts were advocated by many natural philosophers in the early seventeenth century, including Isaac Beeckman (1588–1637) and Galileo. Thomas Hobbes (1588–1697) courted ecclesiastical unpopularity as much for the comprehensiveness of his Atomistic explanations of both natural and human phenomena as for his political and religious unorthodoxy. Another, less radical, reaction to the decay of Aristotelian cosmology was represented by René Descartes (1596–1650). Descartes' universe, like Aristotle's, was an immense plenum. Aristotle's primary matter was replaced by a Cartesian prime matter which is true substance but has no forms or qualities other than extension and motion. This was a radical break with scholastic matter-theory since according to that view quantity and motion were accidental forms and could not confer substantiality on anything. In the Cartesian cosmology the matter of the universe was divided into three principal elements — a very fine material pervading everything, in rapid motion, and preventing the formation of any void. It is of this material that the Sun and the fixed stars are composed. Larger but still 'subtle' round particles form the sky and the vehicle of light; and relatively larger particles of gross matter form the Earth, the planets and the comets. All matter is in principle divisible and moveable and there are no atoms and no void. All motion involves the displacement and the replacement of matter and therefore requires a circulation in the finer parts of matter, the origin of the vortices.

Many thinkers — and not only the Cartesians and the Peripatetics (that is the Aristotelians) — found the concept of a void philosophically unacceptable and felt the need to fill all space unoccupied by manifest bodies with an invisible 'subtle matter' or aether of some kind. During the second half of the seventeenth century, when Newton conducted his theoretical and experimental investigations, the theory of an aethereal medium and of the particulate nature of matter was widely accepted by most scientists, and Newton was no exception. However, with Newton, and strongly influenced in this respect by Isaac Barrow (1630–77), we find an explicit awareness of the difference in the certitude of the reasonings based on the new postulational and inductive methods and the 'fancying' of the earlier philosophers of nature, still strongly represented by the Cartesians. For Newton an hypothesis merely offered a conjectural or possible explanation uncontrolled by experimental testing and had no place in a rigorous mathematical natural philosophy. Newton did hypothesise freely but clearly marked the inferior cognitive status of speculations, and admitted without embarrassment a considerable evolution in his hypotheses. He discussed even more tentative hypotheses by using the device of putting to himself a query and offering a speculative answer. Newton thereby introduced considerable epistemological rigour into scientific reasoning.

Newton's hypothesis of the structure of matter was that it was a hierarchy of particles of different orders, but ultimately composed of indivisible solid particles: '. . . we concluded that the least particles of all bodies to be . . . extended, . . . hard and impenetrable, and moveable and endowed with their proper inertia. And this is the foundation of all philosophy'. Newton also supposed attractive and repulsive forces between these particles and used these forces to explain chemical action. He accepted the possible existence of void between the particles and between planets.

Dynamism

Gottfried W. Leibniz (1640–1716) constructed a highly original and very metaphysical physics from fragments of scholastic and Cartesian philosophy. His thinking is complex and sometimes obscure and only a few of his notions are presented here to outline his theory of the ultimate structure of matter. Leibniz was not satisfied with the Cartesian reduction of underlying matter to extension alone and felt it necessary to reinstate some kind of substantial form. The substantial form which confers reality on matter is for Leibniz 'force'. There are both passive and active forces. Passive force is responsible for impenetrability and inertia, active force causes and confers reality on motion, and is analogous to the soul of living beings. Active and passive forces exist at two levels: as permanent 'primitive forces' underlying all physical change and as 'derivative forces' which, in the case of active force, is a 'conatus' or tendency to determinate motion and enters into dynamics as the cause of particular changes in the world.

Leibniz believed that gross bodies are not true substances but aggregates of simple substances which he called 'monads'. Each monad is a complete and self-sufficient being which does not interact with the others, 'a real and animated point' with its own internal principles of action. 'They all have perception [that is reflect the structure of the world] . . . and appetite [a tendency to change]', and the harmony which exists between these completely independent units is 'one of the most beautiful proofs of the necessity of one substance, sovereign cause of all things, that can be given'.

Leibniz was, of course, aware of the metaphysical character of all of this and he accepted an empirical structuring in natural bodies manifested in division, for example. He employed his axiom that 'nature never acts by leaps' to deny the existence of atoms and the void, referring to them as 'chimeras' without, apparently, noticing the irony of doing so. Leibniz's substantialising of force was to have a considerable influence on Boscovich in the following century and later on the origins of the concept of a 'field', but, as we have remarked earlier, it made the concept of force obscure and distasteful to many positivistically inclined minds.

Was Light Material?

The seventeenth century inherited from Antiquity, and from the Middle Ages, a great variety of theories of light.

According to the atomic theory of Leucippus, Democritus and Lucretius, material replicas issue in all directions from visible bodies and enter the eye of the observer to produce visual sensation. For Plato the process of vision is much more complex involving an emanation of visual 'fire' from the eye and its coalescence with daylight. The homogeneous body thus formed is a material medium through which motions are transmitted to the soul from the visible object. Neo-Platonic medieval light-metaphysics has already been mentioned in Chapter 3. For Aristotle light was not a corpuscular emanation, rather it was the instantaneous actualisation of the potential transparency of the medium between object and observer — not the simplest of concepts. This first determination of the medium is not yet sufficient for vision. Colour, which overlies the surface of the visible object, moves the transparent medium and this acts upon the sense organ producing vision. The Aristotelian tradition, heavily influenced by Islamic and medieval discussions, came to dominate the Latin West in the later Middle Ages and in the Renaissance.

Descartes accepted Aristotle's doctrine of instantaneous propagation but otherwise his theory of light was entirely different and uncompromisingly mechanical, although light itself was not material. Light for Descartes was an instantaneous 'pression' or percussion through the contiguous particles of second substance from a body of the first substance — e.g. the Sun. Light in the luminous body he held to be a 'very prompt and very violent action'. Robert Hooke (1650–1703) in his *Micrographia* (1665) commited himself to the Cartesian hypothesis of an aethereal medium serving as the vehicle of light. Hooke, however, thought of light as undulations in the medium set up by motions in the luminous body. The motion is propagated in every direction with equal finite velocity and propagates in expanding spheres. All colours are the result of the disturbance of the wave-front at refraction.

When Christiaan Huygens (1629–95) first communicated his *Traité de la Lumière* to the Académie Royale in 1679, Ole Roemer had already, in 1676, demonstrated the finite speed of light. This Huygens accepted. He therefore had to endow the Cartesian contiguous aetherial particles with elasticity to account for it. Each point of the luminous body he regarded as the source of its own spherical wave generated by 'the percussions at the centre of these waves'. Huygens' chief contributions to the wave theory of light was his introduction of the concept of a secondary wave propagated from each point of a wave-front, his geometrisation of the wave theory and its application to reflection and refraction.

Newton's views on the ultimate nature of light belonged to the realm of hypothesis and he modified them in various ways. In an early paper, concerning the scholastic issue as to whether light was substance or quality, Newton explicitly decided in favour of substance. His belief that light did not diffract (that is bend around corners) led him to reject the notion that light is a wave motion. He was uninfluenced in this respect by the publication of F. Grimaldi (1613–63) in 1665 of descriptions of diffraction phenomena. Newton's hypothesis of the nature of light — which he strenuously attempted to separate from his experimental findings concerning the heterogeneity of white light — was that light was fundamentally corpuscular. He allowed the corpuscles to excite in the aether swift vibrations which accounted for the 'fits' of easy transmission and of easy reflection which he introduced to explain how a single surface could both transmit and reflect light. Newton's corpuscular hypothesis concerning the nature of light was widely accepted until the end of the eighteenth century. On this view light (and later it came to be thought the same of heat) was a separate kind of matter.

Chemical Theories of Matter in the Seventeenth Century

Chemical praxis in the seventeenth century is characterised by a more detailed empirical study and classification of simple chemical phenomena and substances, and by the beginning of experiment and induction applied to developing theories to explain certain phenomena. Alchemy continued to thrive. In the first half of the seventeenth century four main competing theories were current to provide ultimate explanations for chemical phenomena: the Peripatetic doctrine of the four elements which, in its chemical version, sometimes held that the 'minima naturalia' were themselves aggregates of independently subsisting particles rather than substantial unions of them; the Paracelsian or spagyrist doctrine of the three 'principia', salt, sulphur, and mercury; Cartesianism; and Greek Atomism chiefly revived and 'purified' by Gassendi.

The first serious improvements in the method of a chemical analysis of the nature of matter were made by Johann van Helmont (1577–1644). He made good use of the balance and demonstrated the conservation of matter in chemical processes. He also made the first systematic study of different kinds of gases and indeed coined the name 'gas' from the Greek *chaos*. His theory of chemical compounds was a radical variant of spagyrism. He held that the ultimate inert constituent of matter was water. The active principle which disposed the water and constructed the specific concrete thing was a 'ferment or seminal beginning', an 'archeus' which was generated in matter by Divine light. This brought the 'archeus' to construct the 'seed', which developed into a stone, metal, plant or animal.

Combustion was also studied in the early seventeenth century in the framework of the Paracelsian theory. According to the latter, combustion involved the decomposition of compound substances with the loss of an inflammable oily principle present in the sulphur. Burning would thus result in an increase in weight. Jean Ret in 1630 drew attention to the well known increase in weight during calcination and argued that it could have come only from the air. This presented difficulties for the 'oily' principle (which was later to be developed as the 'phlogiston' theory), which like all theories can be modified and improved to take account of objections.

These older Peripatetic and spagyrist theories were largely banished from chemistry in the second half of the seventeenth century mainly by the lengthy arguments of Robert Boyle (1626–91). Boyle's ambition was to bring chemistry within the scope of the corpuscular and the mechanical philosophy. Boyle's corpuscularianism was similar to the doctrine of the ancient Atomists but, like the Atomism of Gassendi, it was modified to remove its atheistic and materialistic implications. Matter, divided by God at the Creation into particles and motion, is the sole scientific explanatory principle. He imagined small atoms to be combined into minute clusters, the primary concretions which nature very seldom decomposes. These primary concretions form the elements from which compound bodies originate. This theory, similar to that of Daniel Sennert (1572-1637), fixed the notion of chemical elements. In a compound different primary elements may subsist unchanged or they may interpenetrate and demolish each other giving rise to new primary concretions.

Such a theory of course favoured alchemical transformation just as strongly as the older theories, and indeed Boyle and Newton were very active alchemists. Boyle studied combustion in considerable detail and attempted to explain combustion, calcination and respiration as phenomena conditioned by something taken out of the air during these processes. Theorising in chemistry in the seventeenth century rarely transcended what Newton called hypothesis. Towards the end of the century however, as we shall see, chemical hypotheses began to be increasingly controlled by experimental tests of consequences deduced from them.

The Attractive Power of Matter: Magnetism, Electricity and Gravitation

Magnetism

Magnetism was a difficult problem for Peripatetic philosophy, but fitted in much more easily with neo-Platonism and the more occult philosophies. Averroes (1126–98) explained magnetic attraction as a form of multiplication of species, which he conceived as a transmission of the 'species magnetical' from the lodestone, through the intervening material medium. When it reaches the iron it produces in it a motive virtue. In a more familiar Aristotelian vein Petrus Peregrinus argued that the lodestone is an active agent which assimilates the passive iron to itself by actualising its magnetic potency.

William Gilbert (1540–1603), physician to Queen Elizabeth, provided an explanation of magnetic attraction ('coition') and compass orientation ('verticity') which can best be described as strongly neo-Platonic. Gilbert's theory was a consequence of his reflections on his most important magnetic discovery, that the Earth is a great lodestone. Each celestial globe has a specific animate prime form which keeps and orders its own globe internally and externally in relation to the heavens. In particular the Earth has its proper prime form which Gilbert describes as the true magnetic energy. True earth matter (that is lodestone) is magnetic and the prime form is the cause of the Earth's cohesion and of the concordant motions of its matter. Flowing immediately from this form the lodestone has prime powers, much as the human soul has faculties. The chief powers are coition (that is coming together) and verticity (that is the power to orient to the poles). Although each lodestone has in part the power to move itself towards other lodestones these powers are incomplete and do not initiate their motions spontaneously nor do they direct it. Every lodestone is animate and it imitates the human soul. It effuses a magnetic form which, when it inheres in another, completes and specifies its potencies, activates its prime powers and directs that lodestone to a motion of mutual concordance and union.

This explanation differs from that of Averroes in that it is animistic: both lodestones are actively involved in the motion produced, and furthermore no quality or physical state is transmitted through the material medium between the lodestones. Gilbert says that there is an 'orb of virtue' (power) surrounding each lodestone within which the two lodestones can interact and outside of which they cannot do so. The mutual action of one lodestone on another in Gilbert's theory is akin to the coordinating action of the human soul on the parts of its body.

For Gilbert the prime magnetic form of the Earth is responsible for its daily rotation and for the fixity of its axis in space. He made similar prime forms in the Sun and in the planets responsible for the ordered motions of the cosmos, without actually calling them magnetic. He was strongly opposed to cosmic sympathies and antipathies clearly preferring scholastic notions to Hermeticism and the occult, although he vehemently attacked the Peripatetics also, despite being thoroughly Peripatetic in many respects himself. In the course of the seventeenth century various scholastic authors attempted to give a more rigorous Aristotelian explanation of magnetism. However, the interest of the scientific community was largely directed to the empirical results of Gilbert's work, to his concept of the Earth as a magnet and to providing corpuscular explanations of magnetism.

Electricity

Gilbert also laid the foundations of the experimental study of electricity. His style of investigation was not Archimedean but empirical and inductive. His mathematics was descriptive and mensurational rather than deductive. He greatly extended the number of known electrics, that is of bodies which when rubbed attract like amber and jet. His list included diamond, sapphire, mastic and sealing wax. He also described how many objects, when rubbed, do not exhibit electrical phenomena and therefore he distinguished a class of electrics and non-electrics. He also invented a simple electroscope, the 'electric versorium', by means of which he could test the presence of electrification. Gilbert sharply distinguished electricity from magnetism. He explained electrical attraction by a material effluvium because he noticed that materials interposed between the attracting bodies caused a weakening of the attraction, something not observed with magnetism. This would not have occured if the intervening bodies had been properly insulated. Gilbert's 'effluvium' explanation of electricity became widely accepted in the seventeenth century — indeed he has been called the 'father of electrical science'. He denied that any electrical repulsion occurred. Niccolo Cabeo (1585–1650) was perhaps the first to notice that electrics sometimes repel, but it was not until the following century that this was properly distinguished from attraction.

Gilbert's *De Magnete* was a model, not yet perfected, of a new style of natural inquiry which became known as the 'experimental philosophy' in the seventeenth century. He initiated a tradition which is represented by investigators such as Bacon, Mersenne, Mariotte, Boyle, Hooke, Franklin and, most famously perhaps, by Faraday. After Gilbert, magnetic attraction was removed from the realm of a somewhat marginal phenomenon associated with magic. It became a respectable natural force of perhaps cosmic significance, and its scientific investigation a model for explaining other phenomena.

Gravity

By the beginning of the seventeenth century few natural philosophers accepted the existence of the crystalline spheres although many thought there must be a subtle aether throughout space. As an immediate consequence a new explanation had to be found for the motion of the planets. Galileo's belief that circular motion was natural and required no causal explanation satisfied very few. J. Kepler (1571–1630), court mathematicus at Prague, accepted the theory of an innate resistance to motion of all bodies, celestial and terrestrial, and so an external moving agency had to be found. Kepler considered that Gilbert had validly established the magnetic nature of the Earth and he extended this property to all the primary planets.

The rotating Sun impresses a local motion on all of the planets, by a power which is similar to magnetic action. This power acts, for

Kepler, in inverse ratio of the distance between these heavenly bodies. Just as in magnets, each planet has an 'amicable part' attracted by the Sun and an 'inimical part' repelled by the Sun. These secondary forces, superimposed on the tangential driving force of the Sun, explains the ellipticity and variable velocity of planetary orbits.

Descartes adhered to the principle of linear inertia and was aware that an inwards directed tension is necessary to prevent a stone in a sling from flying off at a tangent. In his theory that planets are carried around in a vortex of subtle matter, Descartes located the needed inwards action in a pressure differential in the vortex. As far as terrestrial gravity and magnetism were concerned Descartes explained them also by the motion of corpuscles.

In neither the Aristotelian nor the Platonic tradition was the efficient cause of gravity thought of as an attraction. In the Platonic conception, accepted for example by Copernicus, gravity was thought of as a local tendency for like bodies mutually to unite and it was regarded as a local property of all heavenly bodies. In the Aristotelian tradition it was not a mutual tendency but an innate tendency of heavy bodies to move to the centre of the Universe. Furthermore, it was a property only of two kinds of sublunar matter. Kepler moved away somewhat from the concept of gravity as an innate urge towards the notion that it was an external and mutual attraction. He also held that gravity extends from the Earth to the Moon and reciprocally from the Moon to the Earth, causing tides. Kepler, however, was somewhat ambiguous as to whether gravity was an attraction or an innate urge.

Giles Roberval (1602–75), in 1636, appears to have been the first to publish clearly the notion that terrestrial gravity was a mutual attraction of bodies upon each other, and also that the parts of the fluid matter, which he held fills the Universe, attract each other reciprocally. Gassendi was even clearer. Writing in 1641 he says that gravity is not so much an innate force as a force impressed by the attraction of the Earth. He held that gravity extends to the planets, growing weaker with distance, but not to the fixed stars. Descartes regarded attraction as an animist doctrine, as indeed it was in the hands of Gilbert and Kepler. The widespread acceptance of Cartesianism with its severe mechanical reductionism cast a shadow in the late seventeenth century over attraction as an acceptable explanation, with its animist connotations and suggestion of action at a distance, all of which came under the general anathema applied to occult philosophy and scholasticism.

Robert Hooke (1635–1703), however, between 1666 and 1670 cast doubts on Descartes' vortices as capable of explaining planetary motion and suggested instead that the centripetal deviation of the planets from straight lines was due to an attraction by the Sun. After the publication of Huygens *Horologium Oscillatorium* in 1673, which contained the kinematic laws of centrifugal force, Hooke, Halley and Wren discovered that the law of interplanetary attraction must be an inverse-square law, at least for circular orbits. With all this background which Newton was aware of, what was really original in Newton's astronomical theory, first published in the *Principia* in 1686? There was a great deal that was new, and it is proposed to touch here only on some of those aspects which concern the theory of matter.

Newton showed that the acceleration of a body falling on the Earth was in proportion to the (centripetal) acceleration of the Moon towards the Earth as the inverse square of their distances to the centre of the Earth. This was demonstrative proof of Newton that terrestrial gravity extends to the Moon and beyond, is responsible for the circularity of the Moon's motion and decreases in inverse ratio with the square of the distance. By applying widely accepted rules of inductive reasoning, and some of his own, Newton made the gravitational force exerted by the Sun on the planets responsible for their elliptic orbits and proved mathematically that a centripetal force, varying as the inverse square of the distance, could indeed produce elliptic orbits. Newton therefore made gravity a mutual attractive force between all bodies in the Universe. With his mathematical theory of gravity he was able to explain the tides, and certain anomalies in the lunar motion, with great rigour and conviction. He thereby set in motion a programme of astronomical research which was to be enormously successful for the next two centuries.

Newton was attacked by the Cartesians and by Leibniz for his theory of gravity as an attractive force and was accused of reintroducing occult qualities. Newton responded in effect that his use of the word 'attraction' did not imply any particular mechanism for gravity, and that he was only concerned with establishing that such forces do actually exist, and with the mathematical laws they obeyed. This explanatory austerity was to have a powerful methodological influence in the following century, but in Newton's day the occult connotations of the word 'attraction' were too powerful to be easily ignored. Furthermore, however he might protest, the word attraction implies at the very least the action of one body upon another at a distance as in magnetism. It was ultimately in this form that Newton's law of gravity was thought of in the eighteenth century.

Throughout the seventeenth century the dominant theory of matter was atomic or corpuscular, and the most carefully studied properties of matter were mechanical. But Leibniz kept alive a more subtle theory, according to which the atoms were themselves the products of systems of forces which had a more fundamental being. Eventually it was to be the mechanically puzzling attractions (and repulsions) of magnetism and electricity, and the apparent universality of gravitational attraction that were to tip the scale in favour of matter theories more or less on the Leibnizian model.

Further Reading

R. Dugas, *Mechanics in the Seventeenth Century*, transl. F. Jacquot (Neuchatel, 1958)
A.R. Hall, *From Galileo to Newton* (London, 1963)
D. Lindberg, *Theories of Vision from Al-Kinde to Kepler* (University of Chicago Press, 1976)
J.R. Partington, *A History of Chemistry*, 4 vols. (London, 1961–70)
A. Sabro, *Theories of Light from Descartes to Newton* (London, 1967)
F. Yates, *Giordano Bruno and the Hermetic Tradition* (London, 1964)

5 EARLY-MODERN ART, 'PRACTICAL' MATHEMATICS AND MATTER-THEORY

Pio Rattansi

I

As a Cambridge undergraduate in the early 1660s, Isaac Newton became acquainted with two radically different conceptions of matter, forming part of two rival world-pictures. One of them was to be found in the neo-scholastic textbooks prescribed by his tutors and formed the basis of the Arts course. Newton discovered the other conception in the writings of recent works of the new 'mechanical' natural philosophers.

Ultimately Aristotelian, the scholastic conception regarded matter as knowable only by analogy. Matter was purely potential by itself and desired form 'as the female desires the male and the ugly the beautiful . . .'. Only in thought was matter separable from form since, in fact, whatever had matter also possessed form. Form was not merely shape, but the inner nature of an entity as expressed in the plan of its structure. Was matter a 'substance' — that is a self-subsistent entity? Imagine a sensible body stripped of all that was not substance. Length, breadth, and depth too, must be thought away, since 'quantity is not a substance'. Would matter alone then remain? What was left would neither consistute a certain thing, not be of a certain quantity. None of the categories which determined Being could be applied to it. It would be a 'pure potentiality'. Yet it was clear that the hallmarks of substance were separability and 'thisness'. So it was the addition of form, and the resuling compound of form and matter, that constituted substance. It could not be matter alone, far less any mere combination of matter and shape.

Change as a process could be made intelligible, according to Aristotle, only through the correlative concepts of form and matter, together with privation or potentiality. The naïve view of matter as self-subsistent 'stuff' was inadmissible, as was even the more sophisticated matter-theory of the Atomists.

If a body was divided through and through, what remained would either be nothing or merely a collection of mathematical points, from adding however many of which a sensible magnitude would never result. Matter was potentially divisible into its smallest parts, but was not actually so divided. The Atomists were wrong, moreover, because they derived the properties of an entity merely from matter and its arrangement. Letters and juxtaposition did not make up a syllable any more than bricks and their arrangement a house. A form was always necessary. Nor could any genuine alteration of substance result merely from association or dissociation of particles. It was true that water changed into air more readily if it was already dissociated into small droplets; but a substantial change required a genuine replacement of form.

The role of mathematics in the study of nature was greatly diminished by such a conception of form and matter. Aristotle pointed out that some existing entities were like 'snub' and others like 'concave': 'And these are different because ''snub'' is bound up with matter (for what is snub is a concave nose) while concavity is independent of perceptible matter.' The physicist was concerned with natural things, and they were analogous to the 'snub', having forms separable in thought but never really existing apart from matter. The mathematician dealt with sensibles but not as sensibles; from a snub-nose, he abstracted the 'concave curve'. Dealing with simpler entities he succeeded in achieving far greater precision because 'a science which abstracts from spatial magnitude is more precise than one which takes it into account . . .'. Optics, harmonics, and mechanics similarly, considered objects *qua* lines and numbers but not in relation to sight, sounds or locomotion — even though those attributes were 'proper' to those objects. Physics studied a world of Becoming. The four Elements, capable of mutual transmutation, were far more appropriate to explaining such a world than the static and timeless forms of mathematics. Numbers and spatial magnitudes could legitimately be studied apart from things, but they never really existed apart from them. The objects of mathematics could not, therefore, provide the primary principles by which to explain nature.[1]

Newton found a quite different conception of matter in more recent thinkers.[2] When Descartes tried in his imagination to deprive a sensible object of all attributes that did not truly belong to it, he was left with nothing but 'a certain extended thing which is flexible and movable'. All other qualities could then be derived from extension alone. Others, notably Gassendi (and the English thinkers Boyle and Charleton who followed him), added impenetrability to extension as jointly essential to define matter. All, however, agreed in making matter a self-subsistent entity and rejecting the necessity of form as an immaterial and immanent principle, 'a kind of soul' (in Boyle's words) 'which, united to the gross matter, comprises with it a natural body'. Matter was not merely divisible, but actually divided into subsensible particles. Their shape, size, motion and mutual relations of position and order enabled them to act on each other and on human senses. Boyle said almost all qualities seemed 'mechanically producible', resulting from the size, figure and contrivance of parts. According to Robert Hooke, the Royal Society felt it had 'reason to suspect' that qualities were 'the same products of *Motion, Figure*, and *Magnitude*'

As long as changes were explained by what Hooke now castigated as 'two general and (unless further explicated) useless words of *Matter* and *Form*', mechanics — in common with mathematics — was considered as having little importance for physics. Mechanics was concerned with the quantitative analysis of locomotion, but quantity was not related to essence. With the revised idea of matter, mechanics became the core of physics. To be 'material' was to possess those properties which brought a body within the descriptive and explanatory scope of mechanics. For the first time it was possible to envisage the possibility of a mathematical *science* of nature. Descartes sketched one in his *Principia Philosophiae* (1644) — if in a disappointingly 'hypothetical' and qualitative fashion. It remained for Newton, who as an undergraduate soon began to reject Peripatetic notions and to accept the existence of indivisible atoms, to realise the dream magnificently in his *Mathematical Principles of Natural Philosophy* (1687).

II

What brought about such a momentous change? 'Internalist' accounts point to various pressures which finally resulted in the clear emergence of the notion of 'mass' in Newton's work. Some of the pressures were at work within the medieval Aristotelian tradition. Condensation and rarefaction were important phenomena in Aristotle's physics. The continuing movement of a projectile after it lost contact with the projector came to be explained by an 'impetus' imprinted on the projectile. Understanding how the 'accidents' of bread and wine could be preserved in the Eucharist when the substance was not was an important problem in a theological milieu. Each of these problems seemed to demand the assumption of some invariant quantity in physical bodies.[3]

Other pressures originated in discussions of mixtion. When two substances came together to form a compound, in which even the eye of a lynx could detect no trace of the original constituents, the compound resulted (according to Aristotle) from the constituents acting on each other through their contrarieties. That was easier when their smallest parts had such a shape that they could most easily be juxtaposed. These *minima naturalia* came to be considered by medieval followers of the Islamic commentator Averroes not as limits of logical divisibility, but as entities with an independent existence. Aristotle's explanation had been directed in part at the Atomists, denying that bodies could actually be divided into 'least' parts, or that combination could be reduced to 'composition'.

Forms were assigned the primary role in change: the minima of various substances differed qualitatively, possessed a characteristic size and had to act on each other to render themselves capable of receiving a new form. But by the late fifteenth century there was a pronounced tendency to visualise or 'physicalise' minima, turning them into solid particles and giving prominence to their quantitative features. Such physicalisation of concepts is noticeable in many other scientific contexts besides the minima by the sixteenth century. So, too, were the Galenic physiological 'spirits' in the human body, the great variety of spirits which chemists succeeded in extracting from natural substances, and the mineral or lapidific spirits of mineralogy. The spheres which were thought to carry around celestial bodies came to be physicalised and were expected to shatter like glass if comets were truly above the moon and could crash through them. It is not an exaggeration to speak of a pervasive 'conceptual degeneration' in sixteenth-century natural philosophy.[4]

'Internalist' accounts do little to explain why these various pressures should have resulted in the replacement of one conception of matter by a very different one. Certainly, a change in the fundamental categories for understanding and explaining natural phenomena had occurred by the mid-seventeenth century. That is clear from the works of more recent thinkers which Newton consulted as an undergraduate. Artistotle had declared that the precision of mathematics was unattainable in physics, which dealt with the complexities of a compound of form *and* matter. Descartes, however, made ideas approaching mathematical ones in 'clarity and distinctness' the hallmarks of acceptability in physics.

Those criteria led him to reject all other sensory qualities as confused and obscure or as dependent on the primary characteristic of extension. Artistotle had pointed out the difficulties of identifying the mathematical and the physical, notably that of finding an analogue for the geometer's dimensionless points from which he successively built up lines, surfaces and solids. These difficulties did not deter Galileo from accepting that identification in his early *de motu (c.* 1590), although in reverting to the problem in the great *Dialogue* and his final *Discorsi*, he admitted that the philosophical perplexities lay 'far beyond our grasp'.

Others claimed that sense-experience immediately disclosed extension and impenetrability jointly as essential to corporeity, but they were then confronted with the problem of inferring the nature of subsensible particles from what experience revealed to us about gross bodies. A repudiation of the traditional concept of form and matter, substituting the idea of bodies made of matter, a self-subsistent stuff, and retaining the idea of form only if it signified size, shape and configuration — these were basic assumptions in the 'neoteric' authors whom Newton studied. It was a revision and replacement of fundamental metaphysical categories rather than any scientific disproof of Aristotelian science through inductive inference from observation and experiment. While the scale and character of the metaphysical shift has been analysed with great subtlety, we still lack any convincing account of the purely internal dialectic which is said to have brought it about.[5]

Nor have 'externalist' accounts contributed much in the past to tracing the roots of so momentous a transformation. Mercantile capitalism is said to have involved handling of materials and a calculating and quantitative attitude. With the increasing economic dominance of mercantile capitalists, those habits of thought are said to have become more pervasive and paved the way for the conception of the world as a collection of material bodies, whose nature and qualities could be explained in mathematical and mechanical terms. It has reasonably been objected, though, that there is not even a plausible account of the path leading, say, from double-entry bookkeeping and commercial arithmetic to the mathematical world-picture. Technological needs cannot be said to have been primary in the rediscovery of Greek mathematics or the study and development of astronomy. If the scientific revolution was conceptual and metaphysical, then the ground is cut from under those who wish to assign an important role to technologists for conferring a new significance on experiment. The novel importance gained by experiment appears then a consequence rather than as a cause of the revolution.

III

The weakness of crude externalist accounts are patent. A more sensitive externalist approach ought to examine two areas of activity in the late fifteenth and the sixteenth centuries. One is the history of art, the other the history of practical mathematics and mechanics. They are of interest for charting changes in the concept of matter because the metaphysics of form and matter provided a common framework for description and explanation in these fields, no less than in natural philosophy. Important modifications of that framework seem to have taken place earlier than in the more scientific context. Moreover, it is easier to establish the influence of external factors in these areas. There were changes in patterns of patronage and in style (connected with the taste of the new patrons) in art. The practical mathematicians and mechanists were responding more directly to technical needs than those studying such subjects in a more theoretical and academic milieu. Another external factor was the struggle of both groups of practitioners to raise their own social and intellectual status from the medieval assimilation of their *métier* to that of craftsmen.

It may even be suggested that unless we reconstruct the common context within which discussions we would tend to differentiate as belonging to art, or to practical mathematics, or to science took place in the early-modern period, the true importance and significance of many features of the scientific revolution must escape us. Copernicus, for example, saw the greatest virtue of his new heliostatic system in establishing the 'unchangeable symmetry' of the parts of the universe: 'the heavens themselves become so bound together that nothing in any part thereof could be moved from its place without producing confusion of all the other parts of the Universe as a whole'[6]. Galileo adhered to and gave greater meaning and depth to the Copernican conviction that circular motion was best suited to ensuring and maintaining symmetry and harmony in a well-ordered universe.[7] Such notions have seemed an embarrassment to those concerned to uphold the modernity of Copernicus and Galileo and they have regarded them as a residuum of ancient and medieval ideas from which even those heroic pioneers could not entirely break away.

Yet the idea of beauty as the rational integration of parts, with each part having a fixed size and shape so that nothing could be added or removed without destroying the harmony of the whole, was not a medieval survival but a novel aesthetic ideal given currency in the writings of the painter and architect Leone Battista Alberti, who believed no shape was better adapted to fulfil those demands than the circle and its derivatives. Kepler's speculations on the role of the Platonic polyhedra in the construction of the universe and on the relations between musical harmony and the orbits and angular velocities of planets yield their full meaning — in common with the pronouncements of Copernicus on circular motion — within this artistic context. Moreover, radical changes in the form-matter metaphysics in the context of art seems to have exerted an influence which was not confined to the 'fine arts'.[8]

To use the terms 'art' or 'fine arts' in the earlier period to signify painting, sculpture and architecture is to import our own classification anachronistically into the past. Until the Renaissance, they were denied the intellectual and social prestige and status of the 'liberal arts', since they involved manual work and were therefore grouped with crafts. Alberti was deeply engaged in the struggle to raise the social and intellectual status of the artist. Arithmetic, geometry, astronomy and music had traditionally found places among the liberal arts. Alberti claimed the same status for the fine arts because he said that they, too, were based on mathematics. The systematic use of a new mathematical device in the organisation of visual space in painting was an important Renaissance discovery by Brunelleschi (1377–1466) and was tied to an important stylistic transformation. Already in the late thirteenth century Giotto had departed from the hieratic, symbolical, 'dematerialised' modes of representation of the Byzantine *styla graeca* — mocked by the art historian Vasari in the mid-sixteenth century for its 'wild eyes and bodies poised on tiptoe, and long tapering hands, and that absence of shadows, and the other monstrous conceits of the Greeks'. Instead, Giotto used an 'optical perspective' to paint solid, fully-rounded figures moving in a 'credible' space. But it was Brunelleschi who pioneered an explicitly mathematical perspective, based systematically on central projection of three-dimensional space on a plane.[9]

Art historians trace the emergence of the new stylistic ideal to the increasing importance of lay patronage and a lay piety finding greater satisfaction in Franciscan preaching and dwelling on the human drama of Gospel themes to arouse adoration and pathos, rather than in scholastic theology, with its counterpart in an art disdaining all illusion of depth and relief in two-dimensional representations of Christ *Pantokrator* or the Virgin as *Theotokos*. As it emerged in northern Europe, naturalism continued the tradition of the art of illuminating church windows and manuscript illustration and expressed itself through an accumulation of realistic detail. By contrast, in Italy Giotti relied far more on mathematical devices of composition and the relation between figures and the space enclosed them.

Basing himself on the discovery of mathematical perspective, Alberti offered a justification of naturalism in painting and, simultaneously, of the high calling of the artist. For an adequate appreciation of the novelty of his claims, it is necessary to recall the depreciation of illusionistic painting in ancient and medieval thought. Plato compared the carpenter, realising the ideal form of a bed in wood, with him who made painting of a bed: the painting would be 'an imitation of an imitation' and at a third remove from the 'real'. Plato was particularly scornful of the illusionistic painter who tricked the eye by deforming the true proportions of things. Such painting appealed to that irrational side of ourselves which was deceived by the straight stick which appeared bent in water, took the concave for convex, and was delighted by the conjurer's tricks. How different from the arts of measuring and weighing, which corrected the errors of sense! The mathematician was to be admitted but the painter excluded from the Platonic Republic because he nourished the irrational part of the soul. Even Plotinus, who said that in imitating forms the painter was no different from nature which did likewise, ultimately gave art a lowly place. The beauty achieved by the artist would always be limited by the resistance and inertia of matter to his injection of form into it, and there was a danger of the soul being so entangled in sensual images as to desert its true task of contemplating the ideal forms. Aristotle's view was very different. He admitted no resistance in matter to the imposition of form: indeed, as already noticed, he believed that matter longed to be 'informed' and had the capacity to be informed to the same extent that the form had the power to do so. But nature was for Aristotle an internal moving principle which, in plants, animals and minerals guided their development towards the full possession of the form of their species. Art imitated nature in the sense of copying her *modus operandi*: an internal movement in the soul of the practitioner of the art (who might be physician or farmer, no less than painter, sculptor or housebuilder) was translated into the movement of his hand in order to implant a form in matter. But just as the eternal and immobile forms of mathematics had seemed inept to Aristotle for dealing with the world of Becoming so, too, the static forms of painting and sculpture did not appear to him best fitted to imitate nature. Music was far more suitable, and tragic drama had even richer resources for doing so.[10]

Partly in reaction to the varieties of Gnostic heresy it encountered in its early history and Manichean ones in the thirteenth century, Christian thought from Augustine to Aquinas placed particular emphasis on the beauty of natural and artificial things as a reflection of divine unity through a harmony expressed in number and measure. The most orderly of human senses, those of sight and hearing, responded best to them. But these doctrines were overlaid by others, warning of the dangers of the snare of sensuous beauty unless the sensuous remained no more than a sign and symbol of invisible and unearthly verities and realities. Mimesis continued to be regarded, in accord with the Aristotelian tradition, as an imitation of the working method of Nature, not a faithful depiction of the natural world. Since no essential relation was perceived between the external appearance of an object and its inner form, little spiritual value could be granted to naturalistic representation. The virtues of order and measure were acknowledged, but thought to find realisation in the liberal arts, and, even more so, in theology which, being closer to the supreme source of all forms and of light, was most impregnated with them.[11]

How did Alberti, then, justify the naturalistic ideal of style, transforming mimesis into faithful copying of natural (or artificial) objects, and appeal to the geometric rules of perspective as rules learnt from the theory of vision, enabling the painter to copy nature just as did the eye? Alberti offered a purely phenomenalistic definition of beauty as a harmonious integration of parts, resting on correct proportions to which an inborn sense of harmony in the human mind responded. The architect's understanding of those proportions enabled him to design beautiful temples using the circle and other shapes related to it, thereby echoing celestial harmony and inducing in worshippers feelings particularly pleasing to God. Those proportions existed in natural things and mathematical perspective enabled the painter to copy them faithfully. The visible was implicitly recognised thereby as an embodiment of invisible and intelligible forms through a harmony of proportions. Painting, sculture and architecture could therefore be considered sister-sciences to music.

IV

Alberti's attitude is clearly anti-Aristotelian: beauty was considered a harmony of mathematical proportions and orderliness in the universe, constituted by mathematical relations between visible bodies. Was it a return to Plato? By making the painter a kind of mathematician, it blurred the Platonic contrast between painter and mathematician. But by employing mathematics in the service of illusionism, it was open to the principal Platonic accusation: that 'making' was not to be confused with 'turning a mirror round and round' to reflect all things. The mirror-image was now invested with a positive value. That 'art apes nature' was no longer, as in medieval times, a sign of its inferiority but its chief claim to glory: Alberti praised art for holding up a mirror to nature. Nor can Alberti's opinions be said to be Plotinian, for Plotinus had expressly rejected the definition of beauty as a harmony of parts which Alberti revived and to which he gave deeper content. According to Plotinus the possession of parts was a consequence of the fact that pure forms had to realise themselves in the world of matter, although forms were themselves indivisible. The beauty of a house, for example, was fully revealed only when we thought away the stones to disclose the form. To include the existence of parts, an external fixture unrelated to essence, into the very definition of beauty was quite unacceptable to Plotinus. Alberti's definition and understanding of beauty seems, in sum, to subvert the form-matter metaphysic within which problems we would regard as aesthetic were discussed. Matter, as three-dimensional space-occupying stuff, now gained far greater autonomy and significance than it had been granted in the dominant traditions of ancient and medieval thought. Form primarily signified geometrical shape and size and the mathematical relationship of ratios in a whole. The conception of harmony and of the order which constitutes such harmony — which Copernicus, Galileo and Kepler adopted in their work — was thus that which had been well established in an artistic context through Alberti and his successors by the early sixteenth century; order in the universe had come to signify the most fittingly arranged visual relationship between solid three-dimensional bodies.[12]

Just as the visual and the intelligible tended to become merged through mathematically proportioned harmony in art-theory, so, too, did the physical and the mathematical coalesce in the ideas developed by practical mathematicians and mechanists in the sixteenth century. The two groups of practitioners were not clearly distinguishable, as the term 'artist-engineer' (appropriate to a figure like Leonardo da Vinci) well illustrates. The indispensability of mathematics to his vocation constituted the practical mathematician's chief claim to superiority over mere craftsmen, as it had already been for the artist. Ancient and medieval thought, however, had recognised a clear distinction between practical technological problems and their transformation into the subject-matter of theoretical mechanics, which was itself grouped with the liberal arts. From Aristotle's school originated a celebrated treatise on *Mechanical Questions*, which explained more complex machines in terms of the lever — not the physical lever but an ideal one, made of geometrical line-segments, ideally smooth planes, flexible cords and pullies operating without friction. Such a treatment yielded results which were not in conformity with everyday experience. The distinction between theoretical and practical mechanics continued to be maintained in the theoretically far more consistent and superior work of Archimedes, as also in the late-medieval School of Jordanus. The resulting incompatability with experience was hardly considered a matter for surprise, being attributed either, with Plato, to the recalcitrance of matter, or, with Aristotle, to simplifications inherent in a mathematical treatment which abstracted the 'accident' of quantity from a physical body which was an integral composition of matter *and* form.

These careful distinctions came to be obscured by the practical mathematicians of the sixteenth century. Self-taught for the most part, they served cities and princes, advising on such projects as irrigating fields, draining swamps, making more efficient use of water-power for industrial purposes as well as of the evermore expensive artillery in war, civil and military constructions, or the use of falling weights and stretched strings in time-pieces and planetaria. They presented themselves as the inheritors of the traditions of Classical mechanics and Hellenistic technology. They made Daedalus and Icarus into symbols of the power conferred by mechanics rather than warnings, against overweening human ambition. They presented the image of an Archimedes who gloried in the invention of marvellous mechanical devices, not as the upholder of an ancient distinction, disdaining the application of knowledge to practice. If the ancient and medieval division of theoretical and practical mechanics was to be overcome, the physical bodies of ordinary experience had to be brought far closer to the mathematical bodies. The Peripatetic conception, which made the characteristic qualities and properties of a body or its essence depend on form while regarding its quantitative attributes as accidental, ruled out such a move. A revision of the traditional distinction depended crucially on a new conception of matter, and, in the words of sixteenth-century practical mathematicians on mechanics, we find the claim that what is found to be true in theory is 'verifiable in matter also by the sense of sight' (Tartaglia). Mechanics was declared to be 'no longer . . . mechanics when it is abstracted and separated from machines' (Guido Ubaldo). Whatever helped builders, carriers, farmers, sailors and others came within the scope of mechanics.

The limitations of these writers on mechanics are well recognised. Their achievement was piecemeal rather than systematic for the most part, succeeded only partially in breaking with Peripatetic notions even when it was opposed to them and greatly underestimated the difficulties in passing from a purely mathematical treatment to a physical one. The way ahead lay, as we know, in adopting the principle that every deviation from behaviour according to theoretical principles or laws could be explained by counterforces which could themselves be specified in terms of precise laws. With Galileo and his successors the task was undertaken with intellectual resources and awareness of previous intellectual traditions so much greater as to mark a qualitative transformation. Nevertheless, we must not by any means overlook the indispensable role of the earlier and more practically-inclined men in breaking with the ancient and medieval assumption of an unbridgeable gulf between mathematical and physical, based on the traditional metaphysics of form and matter.[13]

V

In this brief overview, we have speculated on two external influences which may have impinged on matter-theory and contributed to its transformation by the mid-seventeenth century. One is the taste of lay patrons in modifying style in painting in the late medieval period and the Renaissance, the other, the struggle of practitioners, painters, sculptors and architects, as well as practical mathematicians, to raise their intellectual and social prestige. Lay patronage contributed to the adoption of naturalism in painting which was consolidated by the discovery of mathematical perspective in Italy. Naturalism was extended from depiction of predominantly religious themes to glorification of nature and human beauty through a redefinition of beauty as mathematically ordered integration of parts in a whole. The practitioner's mastery and use of mathematical proportions had provided the basis of his claim to equality (or even superiority) to those versed in the liberal arts. The practical mathematicians now followed the artists in merging the mathematical and the physical to claim a similarly exalted intellectual and social status.

The idea of order as a harmony created by positional arrangement regulated by ratios and proportions did not remain confined to the artistic domain. It had important repercussions in cosmology. It provided the justification for Copernicus' violent rejection of the equant as an acceptable device in astronomy and a weighty argument for his own system in the face of all the objections to which it was vulnerable. Galileo continued to place enormous emphasis on the role of circular motion in ensuring the preservation of the initial orderliness of a well-arranged universe. Kepler speculated on the rôle of the Platonic solids and of musical harmony in determining the structure of the cosmic order.

These developments had implications for matter-theory, because they tended to subvert the traditional metaphysics of form and matter. Changeless mathematical aspects, abstracted or 'lifted off' existing things or bodies (in Artistotle's interpretation), were now assigned what he would have considered an inappropriate significance in relation to the essence of a ceaselessly changing world. But the Platonic split between ideal mathematical forms and their realisation in the world was abolished, too. Matter was primarily visualised as space-filled material, with form signifying its purely geometrical characteristics and relations.

External influences appear, in such a conjectural reconstruction, as operating in indirect and complicated ways, interacting with internal practices and traditions in unexpected ways. The practical mathematicians, in seizing upon the tactics deployed by the artists, and blurring the careful traditional contrast between physical and mathematical, undertook a gamble which was vindicated only by the successful efforts of a great succession of masters who carried through that programme. Mathematics and the idea of order were both transformed in the process. The ancients understood number as 'collections of units', and spatial magnitude (expressed as line-segments) as 'determinate numbers of units of measurement'. Number, in the new algebra, was conceived as an abstract and symbolic entity, and the ancient split between arithmetic and geometry was repaired in Descartes' algebraic geometry. 'Mathematisation' was no longer the incorporation of 'apt' ratios and proportions in the positional arrangement of bodies, but an inseparable fusion of mathematical form and physical content. 'Order' came to mean not so much a well arranged integration of parts in a whole, as an orderliness in the working of things and events due to their occurring in accordance with mathematically specifiable laws of nature.[14]

The practice and theory of artists may have been important in giving a new importance and significance to the idea of a mathematical order. But the notion of matter as space-filling material became incorporated by the seventeenth century into the mechanical hypothesis that all change resulted from physical bodies acting on each other at the gross or subsensible level in 'picturable' ways. The conflict which had now emerged between the visual and the mathematical is perhaps nowhere better illustrated than in the initial continental reception of Newton's *Principia*: his mathematical genius was acclaimed but his physics found quite inadequate because it utterly failed to provide any 'picturable' mechanism for the gravitational force whose laws he had formulated and which served to bind together a new 'system of the world'.

Notes

1. Newton's notes in Camb. Univ. Lib. MSS.Add.3996; Aristotle : *Physics* 192a, 194ab; *Metaphysics* 1025b-1026a, 1034ab, 1086a, 1093b; *de gen. et corr.*, 316b-317a.

2. Descartes, *Meditations* (1641), II; Gassendi, *Syntagma Philosophicum* (Lyons, 1658), I, ii, 3; Boyle, 'Origin of Forms and Qualities' (1666) in *Works* (London, 1722), III, 15; Hooke, *Micrographia* (1665), preface; Charleton, *Physiologia Epicuro-Gassendo-Charlotiniana* (London, 1654), I, iii, 1.

3. E. McMullin, ed., *The Concept of Matter in Modern Philosophy* (Notre Dame University Press, Notre Dame, 1963), introduction.

4. Aristotle, *de gen. et corr.*, 327b–328a; A.G. Van Melsen, *From Atomos to Atom*, Eng. tr. (Hamilton, London, 1960), 58–76; E.J. Dijksterhuis, *The Mechanization of the World Picture*, Eng. tr. (Oxford University Press, Oxford, 1961), 200–209; F.S. Taylor, *The Alchemists* (Paladin, St Albans, 1949), ch. 9; R.P. Multhauf, *The Origins of Chemistry* (Oldbourne, London, 1966), 201-236; T.S. Kuhn, 'Boyle and Structural Chemistry in the Seventeenth Century', *Isis*, 43, 1952, 12-36; W.H. Donahue, 'The Solid Planetary Spheres', in R.S. Westman, ed., *The Copernican Achievement* (UCLA Centre, Berkeley, 1975), 144-175.

5. Descartes, *Discourse* (1637) and *Meditations*; II, Galileo, *Discorsi* (1638), Eng. tr. (1914), 44; also I. Leclerc, 'Atomism, Substance, and the Concept of the Body in Seventeenth Century Thought', *Filosofia della Scienza*, 27, 1968; M. Mandelbaum, *Philosophy, Science and Sense-Perception* (John Hopkins Press, Baltimore, 1964), 61-117; E.A. Burtt, *The Metaphysical Foundations of Modern Physical Science* (1932); A. Koyré, *Newtonian Studies* (Chapman, London, 1965) 3-24.

6. Koyré, *op. cit.*, 5-6; M.A. Finocchiaro, *History of Science as Explanation* (Wayne State U.P., Cleveland, 1973), 117-130; B. Barnes, *Scientific Knowledge and Sociological Theory* (Routledge and Kegan Paul, London, 1974), 99-124.

7. Copernicus, *De revolutionibus Orbium coelestium* (1543), Preface; Galileo, *Dialogue Concerning the Two Chief World Systems* (1632), Eng. tr. G. Stillman Drake (University of California Press, Berkeley, 1962), 19, 32, 45, 167, 341; Rheticus, 'Narratio Prima' (1540) in E. Rosen, *Three Copernican Treatises* (Constable, London, 1960), 145; A. Koyré, *The Astronomical Revolution*, Eng. tr. R.E.W. Maddison (Cornell University Press, Ithaca, N.Y., 1973), 58-59.

8. Alberti, *De re aedificatoria* (1485), VI, 2; R. Wittkower, *Architectural Principles in the Age of Humanism* (Warburg Institute, London, 1949), 3-56, 113-116; E, Panofsky, *Idea, A Concept of Art Theory,* Eng. tr. J.J.S. Peake (Ikon, New York, 1975), 48-49; 57-59; J. Gadul, *L.B. Alberti, Universal Man of the Early Renaissance* (1969); A. Koyré, *The Astronomical Revolution,* 119-464; D.P. Walker, 'Kepler's Celestial Music', *J. Warburg & Courtauld Institute*, 10 (1967), 228-250.

9. J. White, *The Birth and Rebirth of Pictorial Space* (Faber, London, 1957); W.M. Ivins, Jr., *Art and Geometry* (1946), 64-86, and *Prints and Visual Communications* (Routledge and Kegan Paul, London, 1953); E.H. Gombrich, *Art and Illusion* (Phaidon, London, 1962); G. Vasari, *Le vite de' più excellenti pittori, scrittori ed achittetori* (1550), 'Proem' to Part II; P.O. Kristeller, 'The Modern System of the Arts', in *Renaissance Thought II* (1965), 163-227.

10. Plato, *Republic*, Book X, 596e-605b; Plotinus, *Enneads,* V, 8, 1; Artistotle, *Metaphysics* 1032b–1034a, *Politics* 1339a, *Problemata* 919b, *de partib. anim.*, 686ab-687ab, *de gen. anim.* 717a, 733ab–736b; Panofsky, *Idea*, 25–32.

11. Panofsky, *Idea*, 35–43; K. Gilbert and H. Kuhn, *A History of Aesthetics* (Thames and Hudson, London, 1956), 119-161.

12. Plotinus, *Enneads,* I, 6; Panofsky, *Idea*, 54–55.

13. Dijksterhuis, *op. cit.*, 241–271; S. Drake and I.E. Drabkin, *Mechanics in Sixteenth Century Italy* (University of Wisconsin Press, Madison, 1969), introd., 3–60, and Tartaglia, 'Questii' (1546) in *ibid.*, 241 and 145; E.W. Strong, *Procedures and Metaphysics* (1936), 47-134; P. Rossi, 'Mechanical Arts and Philosophy in the Sixteenth Century', in *Philosophy Technology, and the Arts in the Early Modern Era,* Eng. tr. (1970), 1-62.

14. J. Klein, *Greek Mathematical Thought and the Origin of Algebra*, Eng. tr. J.W. Smith (Harvard Univeristy Press, Cambridge, Mass., 1968); E. Cassirer, *Substance and Function,* Eng. tr. W.L. and M.L. Swabey (Chicago University Press, Chicago, 1923), 68-76; S. Bochner, *The Role of Mathematics in the Rise of Science* (Princeton University Press, Princeton, N.J., 1966).

Further Reading

General

P.S. Rattansi, 'The Scientific Background' in C.A. Patrides and R.B. Waddington (eds.), *The Age of Milton* (Manchester University Press and Barnes & Noble, New York, 1980), pp. 197-240

On History of Art

E. Panofsky, *Idea, A Concept in Art History* (University of S. Carolina Press, Columbia, 1968)

E.H. Gombrich, *Art and Illusion* (Phaidon Press, London, 1972)

Rudolf Wittkower, *Architectural Principles in the Age of Humanism* (Alec Tiranti, London, 1962)

John White, *The Birth and Rebirth of Pictorial Space* (Faber & Faber, London, 1957)

On History of Mechanics

E.J. Dijksterhuis, *The Mechanization of the World Picture* (Clarendon Press, Oxford, 1961)

S. Drake and I.E. Drabkin (eds.), *Mechanics in Sixteenth Century Italy* (University of Wisconsin Press, Madison-Milwaukee-London, 1969)

E.W. Strong, *Procedures and Metaphysics* (University of California Press, Berkeley, 1936; facsimile by Georg Olm, Hidlesheim, 1966)

6 THE HEROIC AGE OF CHEMISTRY

David Knight

The Proliferation of Fluids

If the seventeenth century was the age of corpuscles, the eighteenth century was surely the age of fluids. By treating the phenomena of heat, light, electricity and magnetism as the effects of exchanges of invisible fluids, physics took if only briefly, a markedly chemical turn.

Phlogiston; the Fluid of Combustion

Chemical theory after Boyle was much concerned with explaining combustion, calcination and respiration. Several English investigators attempted to provide a mechanical explanation for combustion. The traditional alchemical and 'spagyritic' volatile principle occurring in all combustibles was sulphur. Hooke in 1665 regarded combustion as an interaction between the sulphurous particles and air, the supposed universal dissolvent of such bodies. Flame was the site of this interaction which was so violent that it produced a 'pulse of light'. John Mayow (1645–79) thought that nitrous particles were absorbed from the air which 'fermented' with the sulphurous particles during combustion. The theory of combustion that finally triumphed in the early eighteenth century was quite different from those of Boyle and his English contemporaries and was a refinement of spagyritic notions. J.J. Becher (1635–82), a German iatro-chemist in a work of 1669 rejected all traditional elements and principles except water and earth. He distinguished three kinds of earth, one of which, 'oily earth', was contained in all combustible bodies. It is clearly closely related to the earlier 'sulphurous' principle. During combustion oily earth is expelled leaving 'vitreous earth' behind. Becher's ideas were extended by G.E. Stahl (1660–1734) of Halle, who gave vogue to the term 'phlogiston' (from the Greek for 'inflammable') in place of Becher's 'oily earth'. During calcination metals were supposed to give off phlogiston. Ores converted to metal by heating with charcoal absorbed phlogiston from the charcoal. Free air was indeed necessary for combustion, but only to absorb phlogiston. Plants extracted phlogiston from the air, and flames and respiration added it to the air, vitiating it. Combustion and respiration were thus understood as processes mediated by an invisible fluid.

One problem confronting the phlogiston theory was to account for the gains in weight that occurred during calcination of metals. Some chemists supposed that phlogiston had negative weight — a notion close to the old Aristotelian concept of levity — but this idea was not widely accepted. Despite this important difficulty the phlogiston theory gained widespread acceptance after about 1730 and was not superseded until much later in the century due to the work of J. Priestley (1733–1804), C.W. Scheele (1742–86), and above all A. Lavoisier (1743–94), who finally transformed chemistry into an exact quantitative science by exact measurements with the chemical balance.

Caloric: the Fluid of Heat

The theory of heat which eventually predominated in the eighteenth century was that developed by the group of investigators most closely involved with the study of heat: the chemists. W. Homberg (1652–1715) of Paris was one of a group of chemists who made heat and light a chemical element, identified it with the sulphurous principle as one of the primary ingredients in all bodies and present even in the heavens. H. Boerhaave (1668–1738), of Leyden, gave the name of 'fire' to this material, weightless element: an igneous fluid composed of very minute, hard, spherical bodies in constant rapid motion, which penetrates all solid and fluid bodies dilating them. Fire is not simply a solvent of bodies but produces chemical changes also. Boerhaave demonstrated experimentally, weighing metals cold and red hot, that fire has no weight. Heat and light were differentiated when it was found that a plate of glass absorbed heat but transmitted light. Heat or fire, conceived as a chemical element, was even more widely accepted during the second half of the eighteenth century when it became known as 'caloric'. The most important experimental and theoretical work on heat in the mid-eighteenth century was carried out by Joseph Black (1728–99) of Edinburgh. Before Black it was generally thought that equal amounts of heat would raise the temperature of equal masses of any substance by the same degree. Black showed quantitatively that this was not so and developed experimental methods, still employed, for comparing the 'heat capacities' of different substances. He also proved false the belief that when any substance, such as ice, is warmed up to its melting point a small additional amount of heat is sufficient to melt it. He showed in fact that a considerable amount of heat is required for melting and a considerable amount is released on freezing. Black introduced the concept of 'latent heat' in a liquid or vapour to account for this release of heat and devised experimental methods of measuring it. He thus founded the science of calorimetry, but employing a material view of the nature of heat as a kind of fluid in doing so.

The Electrical Fluids

The mechanical explanation of electricity as proposed by Gilbert, in terms of material effluvia, was widely accepted with various modifications until about 1760. This effluvium was supposed to be attached to and to form an atmosphere around electrified bodies. For Gilbert, attraction was due to the tendency of the emanations to return to the parent body; for the Cartesians, it was explained by supposed the effluvia to form vortices, and for W. s'Gravesande (1688–1742), it was supposed to be due to vibrations in the effluvial atmosphere excited by rubbing.

Within this framework of thought a number of important discoveries were made. In 1729 Stephen Gray (d. 1736) found that the electric 'virtue' of an electrified glass tube could be conveyed along lines or wires to bodies as much as 765 feet away. It was found that only a limited class of bodies had this property, to which J. Desauliers (1683–1744), in 1739, gave the name conductors. The electric fluid was included by some among the chemical elements as one of the substances of which the world is constituted. Earlier, in 1733, C. du Fay (1698–1739), as an unexpected outcome of his electrical experiments, discovered that there were two kinds of electrification, each of which attracts electrification of the opposite nature and repels electrification of the same nature.

Independent experiments by W. Watson (1715–87) and B. Franklin (1706–90) in 1746 and 1747 led them both to the conclusion that electrical effects are not due to the production of an effluvium by rubbing but due to an excess or defect of a single electrical aether which can be transferred from one body to another. Franklin was thus the first to introduce the terms positive and negative electricity.

Robert Symmer (*c.* 1707–63) in 1759 put forward a two-fluid theory of electricity. This theory dispensed with the complicated interactions between the electric fluid and the rest of matter implied in Franklin's theory. The two-fluid theory was used by C. Coulomb (1736–1806) in his important researches towards the end of the century in which he showed that the force between two electric charges decreases with the square of their distance apart.

The Magnetic Fluids

Few investigators, except the Jesuits, after the advent of Cartesian mechanism, accepted Gilbert's very scholastic explanations of magnetic actions in terms of powers and virtues. Descartes explained magnetism by postulating screw-like particles issuing from the poles of magnets and screwed threads in iron and magnets, to account for magnetic orientation and attraction. Similar theories were widely accepted beyond the middle of the eighteenth century. Leonhard Euler (1707–83) for example supposed that magnetic materials are honeycombed by fine valve-like pores through which the magnetic fluid could pass in one direction only. However, experimental investigations of magnetism in the eighteenth century were to lead to a non-mechanical theory of magnetic action.

John Michell (1724–93) in 1750 was the first to formulate magnetic action in terms of the forces and distances between magnetic poles rather than between whole magnets. He stated that each pole attracts equally, at equal distances, in every direction, a property difficult to reconcile with the vortex theory. He further stated that magnetic attraction and repulsion are exactly equal to each other and that magnetic force decreases as the inverse square of the distance from the poles. Coulomb was to provide an empirically precise demonstration of the inverse-square law of magnetism between 1785–89.

These discoveries in magnetism, together with the promulgation of a one-fluid theory of electricity, led in 1759 to the construction of a similar theory for magnetism by Aepinus. He supposed the poles to be places at which the single magnetic field exceeds or falls short of its normal quantity. The particles of magnetic fluid repelled each other and attracted those of ordinary matter. The difficulty of explaining the mutual repulsion of negative poles soon led, as in the case of electricity, to the hypothesis of two imponderable magnetic fluids by A. Brugmans (1732–89) and others. The division of a magnet was found always to produce another magnet, never isolated poles, and this led to the supposition by Coulomb and others that magnetic substances were composed of, or contained, tiny magnetic particles, that is particles in which the two fluids could be partially separated but not communicated to other particles.

First Steps in the Chemistry of Gases

In the second half of the eighteenth century, chemists long familiar with the earth and the sea conquered a new region — the air. Minerals had been found to contain different metals, shaking the idea that there was an element 'earth', and in the great age of spas the analysis of different mineral waters was leading towards their synthesis; but the air was still looked upon as an element through it could differ in its general condition, sometimes being purer and sometimes so bad as to be suffocating. In the first half of the century Stephen Hales (1677–1761) collected air from various minerals when they were heated, and found it more or less good; but he did not think of it as composed of various chemical species.

It was Joseph Black (1728-99) in Edinburgh in 1755 who took this step when he worked out what happened when lime ($CaC0_3$) was roasted to give quicklime ($Ca0$), which was then slaked ($Ca(0H)_2$) and slowly returned to lime again by absorbing 'fixed air' ($C0_2$) — so called because it could be fixed in a solid body. He did not of course use formulae like ours. In a series of systematic experiments, he concluded that fixed air behaved like acid; and that it was 'different from common elastic air'. Air then was not an element, but a mixture of various 'airs' as they came to be called.

Black's work was not immediately followed up, but in the 1770s two great experimenters and discoverers came upon the scene. One was C.W. Scheele (1742–86), an apothecary's assistant in Sweden, who first isolated the gases we call oxygen and chlorine. The other was Joseph Priestley (1733–1804), a unitarian theologian, whose hobby was science, and who worked on the chemistry of many gases. His isolation of the substance we now call 'oxygen' was an important stimulus to research, because it was the first to be described in print. The crucial techniques were the collection of gases over water first of all, and then for soluble gases over mercury — which meant that they could be studied directly and even weighed. Black's study had been mostly indirect. Following Priestley's work, Henry Cavendish (1731–1810) established that water was the result of the combination of 'inflammable air' — our hydrogen — with 'eminently respirable air' — our oxygen.

These discoveries raised enormous difficulties of interpretation. As we have seen the accepted theory of burning was that everything that would burn contained something called phlogiston.

When Lavoisier (1743–94) put it beyond doubt that when metals burnt the resulting *calx* weighed more than the metal, this meant that if phlogiston had been given off in the process, it must have negative weight — which was certainly an odd conclusion. On the other hand, many things like coal and candles weighed less and less as they burnt up. While the theory had some anomalies it was not clear that these were serious enough to condemn it completely. Both Priestley and Scheele remained content with it and Humphry Davy (1778–1829) toyed with a revised phlogiston theory in the first decade of the nineteenth century.

Thus it was not obvious to Priestley or to Cavendish that water was a compound; it might be that our oxygen and hydrogen were modified forms of compounds of water, or that hydrogen was phlogiston. Here Lavoisier excelled his contemporaries, and in the 1770s and '80s he put chemistry on to a new foundation by attributing combustion to Priestley's 'eminently respirable air', which was absorbed in burning. Because he thought this gas was present in all acids, he called it 'oxygen'; Lavoisier's name for his discoveries thus enshrines one of his mistakes. For Lavoisier, elements were the limit of analysis; thus any element in any chemical reaction yielded a heavier product since to form a product it had to be combined with something else. In all reactions, one had to take into account gaseous reactants; and when one did, the total weight of the reactants before and afterwards was the same. Chemistry could at last become a quantitative science, with a new and longer list of elements to replace the old Earth, Air, Fire and Water. With a new theory and a new system of names, chemistry was no longer a mere adjunct to pharmacy and mining but a fundamental discipline, concerned with the nature of things.

Atoms and Elements

It was with the emergence of chemistry as a science — and indeed as perhaps the leading science — at the end of the eighteenth century that the prospect arose of an atomic theory of matter which would be testable in detail instead of merely being a generally satisfying but inexact 'picture of the universe'. We associate chemistry with the atomic theory so closely that it is a surprise to find that A.L. Lavoisier, the 'founder' of modern chemistry, should have urged his contemporaries to forget such metaphysical speculations and that Wilhelm Ostwald (1853–1932), the pioneer of physical chemistry, should have opposed Atomism even as late as the beginning of the twentieth century. It was John Dalton (1766–1844) in the opening years of the nineteenth century who revived the atomic theory by making it relevant to chemical analysis. He took seriously what had been for a hundred years a part of the general world-view of men of science, and in making the theory more testable he made some who were previously believers come to doubt Atomism. This is of course the risk faced by any evangelist, Benjamin Franklin (1706–90) was for instance converted to the heretical doctrines of Deism by hearing a sermon preached against them.

In 1793 John Dalton left Kendal for Manchester, where he was to spend the next half century, until his death in 1844. Dalton's time in

Manchester corresponds approximately with the period of greatest intellectual excitement in chemistry. Great audiences flocked to hear the lectures of Fourcroy in Paris, and of Davy and then Faraday in London. It seemed as though chemistry was penetrating below the surface of things and revealing the very structure of matter. New elements were being discovered at a great rate, new agents such as electricity and ultra-violet light were producing new effects and, liberated at last from its role as a service-science in mining and pharmacy, chemistry seemed to be about to solve some of the most ancient and intractable problems concerning matter. The frontiers of chemical knowledge were still close: a bright young man like Humphrey Davy might within a year of taking up the subject be making fundamental studies in it. Having dropped out of a medical apprenticeship, he published in 1800 at the age of twenty-one what became a standard work on the oxides of nitrogen. He was then working at the Pneumatic Institute in Bristol, run by Thomas Beddoes (1760–1808), who had lost his job at Oxford because of his democratic sympathies. The Institution was financed by the Wedgwoods and Davy's friendship with the son of James Watt had got him the post. The Institution was founded to study the application of the gases newly isolated by Joseph Priestley to the cure of diseases, especially tuberculosis, the great scourge of the day. Oxygen and other gases did not turn out to help sufferers much, but Davy discovered the anaesthetic properties of nitrous oxide, and sniffing it became a craze.

His work on the oxides of nitrogen — taking up studies begun by Joseph Priestley — led him to dissent from Lavoisier, whose *Traite élémentaire* appeared in 1789, with an English translation in 1790. There Lavoisier argued against Atomism as throwing up 'indefinite problems', adding that 'if by the terms *elements*, we mean to express those simple and indivisible atoms of which matter is composed, it is extremely probable we know nothing at all about them'. Chemists should therefore come down to earth, leave such speculation behind and go into the laboratory. Then they would 'apply the term *elements*, or *principles of bodies*, to express our idea of the last point which analysis is capable of reaching'. Lavoisier appended to his book a long section on laboratory manipulation, in the expectation that anybody doing experiments his way would come to assent to his theory.

Lavoisier drew up a list of simple substances, or 'elements of bodies', instead of the Earth, Air, Fire and Water then still found in chemistry books, the spagyritic principles, or Boyle's corpuscles. Lavoisier's list included such things as oxygen, hydrogen, sulphur, antimony and manganese — substances which still feature on tables of chemical elements. But he also included light, and caloric — the principle or element of heat — and a collection of radicals, 'oxydable and acidifiable' bases of acids like that from sea-salt, and 'earthy substances' like lime and magnesia. His elements differed from those of most of his predecessors in that they were not omnipresent; on the old view, everything (even ordinary earth or water) contained all three of the principles — sulphur, salt and mercury — in the scheme favoured by the alchemists and followers of the sixteenth-century chemical revolutionary, Paracelsus. Lavoisier's elements were not ideal or perfect substances, but could be isolated in a state of chemical purity. Boyle had proposed a definition not unlike Lavoisier's, but had gone on to reject it in favour of corpuscles differentiating distinctive chemical substances only by their texture or structure. Most of Lavoisier's contemporaries were also corpuscularians, believing that these ultimate particles were all identical to one another although this idea was wholly untestable.

In the older chemistry the elements or principles were also the bearers of properties. Diamonds were predominantly water, and rocks mostly earth; and in the same way phlogiston, the principle of inflammability, was supposed to be present in everything that would burn. Composition determined properties. Although Lavoisier rejected phlogiston, his elements were in some cases similarly the bearers of chemical properties. This caloric, as latent heat, was thought to be a component of all gases. So oxygen and hydrogen were seen as compounds of unknown bases with caloric rather than elements in their own right. More seriously, oxygen was so named from the Greek meaning 'generator of acids' because Lavoisier believed, from the analogy of sulphuric and nitric acids, that all others must contain oxygen too. His successors had to wrestle with Lavoisier's twin principles of a science of matter: that chemists should confine their analysis to the level of elements and not worry about a deeper theory of matter, and that the properties of compounds were determined by the properties of their components. Many, especially in France, Britain and the USA, followed his scepticism; but the problem of composition was soon seen to be more complex than he had supposed.

In 1800 Davy established that the agreeable nitrous oxide, the choking brown fumes of nitrogen peroxide, and the ordinary air that we breathe were all composed of oxygen and nitrogen in different proportions. Important properties did not depend on components, but on proportions and arrangements as the corpuscularians had believed. Davy was also inclined to the older doctrine that heat was motion of particles rather than a kind of chemical element — a notion recently revived by

Count Rumford (1743–1814). In 1807 Davy isolated potassium using an electric battery establishing that the strong alkali caustic potash was an oxide — which would be odd if oxygen were the generator of acids. In 1810 he demonstrated that no oxygen could be found in the dry vapour of the acid of common sea salt (marine or muriatic acid) and that its mysterious basis was not therefore 'oxymuriatic acid gas' but an element, chlorine. Soon he interpreted the substances bromine and iodine as elements similar to chlorine. Since sulphuric acid was believed to be SO_3, it and marine acid, HCl, had no element in common; so acidity could not be the result of a material principle but must simply be due to the arrangement of particles in the compound.

The force preserving the arrangement of particles or corpuscles Davy believed to be electrical; and in his work on electrolysis he believed he had proved this. A chemical reaction could be made to generate an electric current, and conversely an electric current could effect a chemical reaction, such as the liberation of potassium from potash. He found in 1806 that the reactivity of metals depended on their electric charge, so that positively-charged silver was reactive and negatively-charged zinc was inert. Reactivity was not therefore the property of particles of matter, but an effect of forces associated with the ultimate particles. At last light seemed to be being cast on the forces which Newton had exhorted his followers to look for, and the atomic theory began to look rather more definite than it had to Lavoisier. Davy was able to group elements in a tentative way in an electro-chemical series. He used his ideas of reactivity and charge in designing 'protectors' of iron for the navy which when fixed to the less positive copper sheathing used on the hulls of warships saved it from corrosion. These devices had an unfortunate side effect; marine organisms adhered to the protected copper and slowed up the ships; but the same principle of cathodic protection is used today to stop oil-rigs in the North Sea from rusting.

The difficulty about Davy's development of these ideas of corpuscles and electrical forces was that they were not quantitative. Electrical interactions could not be measured unambiguously until Michael Faraday's (1791–1867) work of the 1830s and the thermo-chemistry of G.H. Hess (1802–50) and others in the 1840s. In Paris C.L. Berthollet's (1748–1822) attempts in the opening years of the nineteenth century to understand chemical equilibria following Newton's programme also failed, for the same reasons. A less sophisticated but more workable atomic theory was required. It was supplied by two men in what were then places remote from the great metropolitan centres of science. J.L. Proust (1755–1826), working in Madrid, established against Berthollet that chemical compounds were of constant composition; this is partly a matter merely of definition, but it provided a principle for separating compounds (having constant proportions of constituents) from mixtures and alloys in which the proportions of constituents could vary. John Dalton proposed that the various chemical elements were all characterised by irreducibly different atoms, of different weight for each element. Lavoisier had made chemistry quantitative with his weighings, which he used to overthrow the phlogiston theory, and weight was a quantity easy to measure. Dalton's Atomism led to questions to which more definite answers could be given than did Newton's more sophisticated Atomism. There is a certain irony in this, in that twentieth-century Atomism is closer to that of Newton, Berthollet and Davy than to that of Dalton.

Dalton then required more than twenty different kinds of fundamental particle, against the corpuscularians' one. But given his atoms, if combination was a matter of atoms being linked in simple ratios, then Proust's laws will follow as a matter of course expressing the fact that the same chemical compounds always had the same proportions of elements. Further, if carbon forms two oxides, then there will be twice as much oxygen in one (carbon dioxide) as in the other (carbon monoxide). Dalton assumed that where there was only one compound of two elements, it would contain one atom of each: water was therefore for him HO. With carbon and oxygen, one oxide must be CO, but there was no way of being sure which. If the higher one were CO, the lower must be C_20; otherwise they are CO and CO_2. If hydrogen is assigned an atomic weight of one, then since eight units by weight of oxygen combine with one of hydrogen to give water, the atomic weight of oxygen will be eight. Six units of carbon combine with eight of oxygen in the lower oxide, and twelve in the higher; so, depending on the formulae assigned, the atomic weight of carbon will be three or six. If on the other hand, because one volume of oxygen combines with two of hydrogen to give water we (like Davy) assign a formula to it of H_20, the atomic weight of oxygen will be sixteen, and that of carbon six or twelve.

In the absence, then, of independent ways of determining formulae, Dalton's atomic theory could not bring unambiguous quantification to the theory of matter. Many of his contemporaries preferred to call his atomic weights equivalents, and to see his importance as lying in the confirmation his theory gave to the laws of chemical combination and in the possibility of using equivalents to work out exact chemical recipes. This was indeed the line taken by Davy when as President of the Royal Society in 1826 he presented a Royal Medal to Dalton for his work. Many chemists used the term 'atomic theory' when all that they meant was the laws. Beyond that everything seemed indefinite and speculative. It was not until after 1860 that recognised rules for arriving at atomic weights were agreed upon and that all chemists began to write the same formulae for even the most familiar compounds.

Dalton did not share the belief of many of his contemporaries that the elements were really compound radicals — perhaps, as William Prout (1785–1850) suggested in 1815, all polymers of hydrogen. But his use of the term 'atom' for the ultimate particle not only of elements like hydrogen but also of compounds like water confused his readers. He had also come to chemistry from meteorology, through perplexity as to why the atmosphere was so uniform. Commonsense would suggest it ought to be a sandwich with the densest gas, carbon dioxide, at the bottom, then oxygen and then nitrogen. He and his generation generally saw the particles of gases only vibrating faster as the temperature rose — running on the spot, rather than dashing hither and thither and thus diffusing through space. To explain the constant composition of the atmosphere, most chemists regarded it as a loose compound. Dalton invoked repulsive forces between like particles. His concern with particles and densities led him towards his theory; but his belief in repulsive forces made him reject the possibility of groupings such as H_2.

The great Swedish chemist J.J. Berzelius (1779–1848) systematised the atomism of Dalton and the electrical theory of Davy and, with his support, atomism became a working hypothesis for most chemists. For him, atoms of different elements had characteristic electrical charges, and he saw compounds as composed of positive and negative halves. His theory was therefore called 'dualism', and contrasts with later theories in which the whole molecule was conceived of as a unit. Because all hydrogen atoms, for example, were positively charged, for Berzelius as for Dalton H_2 was an impossible arrangement. Berzelius' equivalent weights were determined with greater accuracy than those of his predecessors. Because they were not generally exact multiples of that of hydrogen, Berzelius was widely held to have disproved Prout's hypothesis of 1815 that all elements were composed of hydrogen. By about 1830, the old Newtonian Atomism seemed to chemists to have been exploded — or anyway to have become irrelevant to their concerns. The debate was between Daltonian Atomists and sceptics, who doubted the viability of any atomic theory.

Arrangements of Atoms

In 1822 in Berzelius' laboratory, Eilhard Mitscherlich (1794–1863) found that many different salts formed exactly similar crystals, and that some substances could form two different kinds of crystal. These phenomena he called 'isomorphism' and 'dimorphism'. They could readily be explained if one assumed that the crystals were composed of atoms. In isomorphism, the same number of different atoms were supposed to be arranged in the same way; while in dimorphism, the same atoms were arranged differently. Those who did not believe in atoms had no alternative explanation to give; but they could point out that salts of the compound radical ammonium were often isomorphous with those of potassium. This implied either that numbers of atoms were not important, or that potassium was like ammonium, really a compound radical (and that other elements might also be complex). Alternatively it could be argued that the laws of isomorphism did not depend on an atomic theory, and that chemists should eschew such metaphysical doctrines and stick to experiments and the laws derived from them.

Talk about the number of atoms in a compound was anyway very difficult because not until 1860 was there agreement about converting equivalent weights to atomic weights, and so different authors gave different formulae to the same substance. Thus potassium nitrate and calcium carbonate are isomorphous; both are even dimorphous, and the two forms of each are isomorphs. This would seem therefore excellent ground for concluding that they contain the same number of atoms. We indeed write them KNO_3 and $CaC0_3$; but the most usual formulae of the 1830s for these two substances were $(K0+N0_5)$ and $(Ca0+C0_2)$, written in halves in the dualistic way. What seemed to have been excellent evidence for atomism turned out to be much vaguer; and crystallography did not yet furnish definite evidence for a structural theory of matter.

Mitscherlich showed that crystalline form was not necessarily characteristic of a substance, and very soon afterwards it was found that chemical composition was not characteristic either. In 1824 Justus Liebig (1803–73) and Friedrich Wöhler (1800–82) analysed fulminic and cyanic acids respectively, and the great French chemist Gay-Lussac (1778–1850) realised that the analyses as proportions of elements were identical. In 1828 Wöhler found that the ammonium salt of cyanic acid turned spontaneously into urea. Later chemists saw this as the first synthesis of an organic compound in a test-tube, but this was not what struck Wöhler and his contemporaries. They were astonished to see one substance turn into another with nothing being emitted; and the only way in which any explanation of what was happening could be given was that in the various compounds the same atoms were arranged differently. The production of urea was then the result of a spontaneous rearrangement.

Berzelius in 1832 named this phenomenon 'isomerism', from the Greek meaning 'equal parts' — isomers being substances of similar composition and dissimilar properties. He distinguished these cases from others where the proportions were the same but one compound contained more atoms than the others. For example butylene, discovered by Faraday, had the same proportions of hydrogen and carbon as ethylene. Both cases seemed to call for an atomic explanation. Attempts were made later in the century to account for isomerism in terms of chemical energy — notably by Benjamin Brodie (1817–80) in the 1860s and by Ostwald as late as 1904 at important public meetings of the Chemical Society of London — but they came to nothing.

The new possibilities of chemical analysis and improved apparatus (especially the condenser we now call Liebig's) for collecting volatile liquids with the hope of determining molecular structures led to a change in the scale of academic chemistry. Berzelius' laboratory had also been his kitchen, and he had supervised one student at a time, like a tradesman with an apprentice. Liebig and others — including Wöhler — persuaded universities to set up teaching laboratories, where numbers of students could work at difficult but relatively straightforward problems for higher degrees. Wöhler, at Gottingen from 1836 to his death in 1882, is supposed to have trained about 8,000 students. The increase in university teaching was matched by an increase in science teaching in schools, as compulsory schooling spread through Europe; and with formal courses at lower levels there seems to have come an increasing dogmatism. Sophisticated experts could make fine distinctions between theories, working hypotheses and systems of classification, but schoolteachers and their pupils were unlikely to do so. The atomic theory, backed with ball-and-wire models which seemed to the Professor of Chemistry at Oxford (Brodie) to be mere materialistic joiner's work, carried the day in the second half of the century.

In France there was a long tradition, reflected in the structure of the Academy of Sciences, in which chemistry was regarded as a branch of natural history. Its chief aim was therefore not explanation but classification and description. Lavoisier's table of simple substances and his remarks about the dangers of atomism could be taken as supporting this view of the science. Attempts by J.B. Dumas (1800–84) and others to fix upon atomic weights ran into great difficulties in the 1830s and proved ambiguous; and the two leading theoretical chemists in France at the mid-century, Auguste Laurent (1808–53) and Charles Gerhardt (1816–56), opted for an essentially taxonomic science aiming merely to classify material substances. Because structures were unknowable — so they believed — the task of the chemist as they saw it was to arrange and classify. Gerhardt was the more ruthless systematiser, and for him chemical equations were merely condensed recipes. For determining atomic weights he drew up conventional rules, believing that certainty was impossible and consistency was all that was attainable. He arranged compounds in series of types, having formal similarities. But although his series look rather like structural formulae, they were only meant to be like Linnaeus' system of botany, an efficient and rule-governed system of codifying knowledge of chemical reactions, not a general theory of matter.

Meanwhile in Britain with the work of A.W. Williamson (1824–1904) on the synthesis of ethers and that of Thomas Graham (1805–69) on the diffusion of gases, there was an interest in the dynamics of chemical change. Davy's work had been also concerned with the forces involved in chemical union, and his pupil Faraday had carried on these studies in his classic quantitative researches in electro-chemistry. Williamson saw chemical reactions as equilibria and followed complex changes through a series of intermediates to the final product, in an investigation which did cast light on the actual structure of ethers where two alkyl radicals, such as methyl CH_3 or ethyl C_2H_5, are linked by an oxygen atom. Edward Frankland (1825–99) at the same time was working on organo-metallic compounds and coming to the conclusion that a given atom of an element could be united with a definite small number of other atoms.

J. Liebig (1803–73) had worked in France, as had Williamson, while Frankland had studied in Germany; and the experience of working in a different tradition was often very valuable in nineteenth-century chemistry. The greatest example of this was F.A. Kekulé (1829–96), who studied in Germany, and then in Paris and in London; and who in 1858 published the paper which laid the foundation of structural organic chemistry. Starting with the idea of the atom of each element having a fixed combining power (later called 'valence' or 'valency'), he conceived of carbon atoms as being able to form chains and believed that information about structures could be obtained from reactions: he claimed to have had a waking dream on the top of a London bus in which atoms grouped themselves in space. In 1864 — again perhaps after a dream in which, as in alchemical symbolism, snakes caught their own tails — he hit upon the idea that benzene was made of a ring of carbon atoms.

Various conclusions followed from this about the number of different substitution products possible. Thus if benzene C_6H_6 is represented by a hexagon with a carbon and hydrogen atom at each corner, then if two hydrogen atoms are replaced by chlorine we get three possibilities:

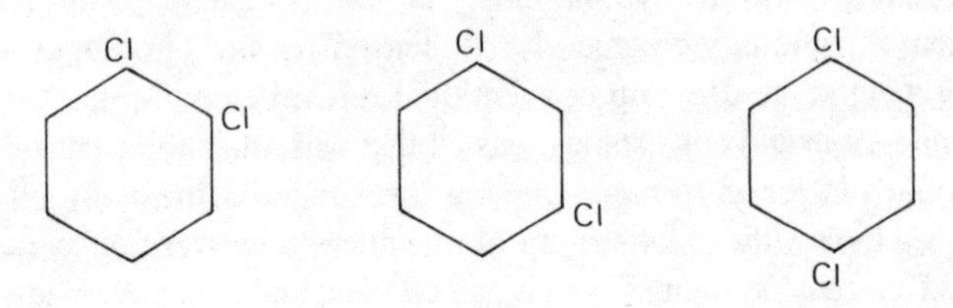

The demonstration that there were three and no more, and the identification of which was which, were made by Kekulé's pupil W. Korner (1839–1925).

Kekulé's earliest diagrams looked more like a mass of sausages than a ring, and he had to invoke single and double bonds in oscillation between the carbon atoms to explain the stability of the ring; but despite these complications, the theory soon caught on, and to study the derivatives of benzene, the 'aromatic compounds', without using the atomic theory became in effect impossible. In 1874, two people who had worked with Kekulé, J.H. van't Hoff (1852–1911) and J.A. le Bel (1847–1930), independently proposed that if the four atoms or radicals were all different then there would be two arrangements which would differ as the left hand differs from the right.

L. Pasteur (1822–95) had found that tartrates crystallise in two forms differing in this way and that solutions of the two forms rotated the plane of polarised light in opposite directions. Now an explanation of this could be given in terms of the arrangement of the atoms in three dimensions. These ideas were extended into inorganic chemistry by Alfred Werner (1866–1919) at Zurich, in researches beginning in the last decade of the nineteenth century. In the twentieth century there came a confirmation of the work of Kekulé and his school, when X-ray studies of graphite indicated hexagons like those proposed for benzene, but for diamonds — the other form of pure carbon — a tetrahedral arrangement like that suggested for the tartrates. Even during the last third of the nineteenth century nobody working in organic chemistry had any real choice about accepting a theory of atoms arranged in space in ways that could be inferred from reactions. Kekulé had no qualms about arranging his carbon atoms into chains or rings because by the 1850s the old dogma that like atoms must repel each other had been abandoned. In 1860 the international contacts that chemists like Kekulé had made individually were formalised with the convening of the first major international conference in chemistry at Karlsruhe. The aim of the conference was to promote consistency in the use of atomic weights; but this was not achieved. The chairmen of the sessions were chosen from the most elderly representatives of each nation, and there was confusion still about atomic weights and equivalents. It was not until they were on their way home that delegates read a paper circulated by the Italian chemist Stanislao Cannizzaro (1826–1910). He urged that the views of his countryman, A. Avogadro (1776–1856), published half a century earlier, which could, if taken seriously, resolve the atomic weight problem. Avogadro had in effect suggested that equal volumes of gases under the same conditions contained equal numbers of molecules and that the elementary gases such as oxygen and hydrogen formed molecules containing at least two, but certainly an even number of atoms.

Thus since water is composed of two volumes of hydrogen and one of oxygen, and volumes represent molecules, it must be that: 2 molecules hydrogen + 1 molecule oxygen = 2 molecules water. Because oxygen and hydrogen never need to be split into more than two, the simplifying assumptions can be made that their molecules are 0_2 and H_2, and the equation becomes: $2H_2 + 0_2 = 2H_20$. The atomic weights then become unambiguous. Oxygen must be sixteen rather than eight times as heavy as hydrogen, and with further reactions unambiguous atomic weights of all elements that form a volatile compound could be determined. Others could be inferred by analogy. Although the Karlsruhe conference had resolved that equivalents were more empirical than atomic weights, and Cannizzaro's paper had not been appreciated there, within a year or two some of the most long-standing objections to the atomic theory in chemistry had been dissolved.

Chemical Classification

It was however still possible to dispute the reality of the atomic structures coming into vogue. The Chemical Society of London debated the issue with some vigour in 1867 and 1869, but in a sense the arguments of the anti-Atomists were the last shots in a war which had, except in the new disciplines of physical chemistry, already been won — rather like the famous battle of New Orleans, fought after the peace had been agreed upon. The irony is that the great chemical achievement following the vindication of Atomism as an explanatory theory was a taxonomic one. If anyone tried to order the chemical elements as Linnaeus had ordered plants, using equivalent weights as a guide as he had used numbers of stamens and pistils, he found that while there were certain regularities and certain families of elements, there was no overall pattern. With the coming of agreed atomic weights, everything changed.

During the 1860s chemists in many countries found that if the elements were written down in order of increasing atomic weight, the list displayed periodicity — that is similar elements came up at regular intervals. The most famous of these chemists was the Russian D.I. Mendeleev (1834–1907), whose Periodic Table first appeared in 1869 and who staked his reputation upon it as none of his contemporaries or predecessors had done. In particular, he left gaps for undiscovered elements. This was an altogether new technique. P.A. Lacepèdes (1756–1825) and William Swainson (1789–1855) had left spaces in their classification of animals for undiscovered creatures, giving some indication as to what they might be like. But Mendeleev's detailed predictions were strikingly confirmed by the discovery of the elements scandium and gallium: by about 1880 his table had come to make sense of an enormous quantity of information in inorganic chemistry and had become a feature of chemistry lecture-rooms. Properties of elements could be inferred from those above, below or beside them, or on a diagonal. The system even proved itself capable of incorporating a whole new family of elements; the rare gases such as argon and neon, discovered by William Ramsey (1852–1916) and others, at the turn of the century.

It is striking how international was the chemical community in the nineteenth century, with fundamental advances often coming from people in relatively remote plates: Dalton in Manchester, Berzelius in Sweden, Cannizzaro in Italy, and Mendeleev in Russia. Although research schools ('centres of excellence') were being built up in universities — especially in Germany — and academies and learned societies flourished in capital cities, major work was often being done outside them. Perhaps this was especially true in chemistry, where metropolitan sophisticates tended to be sceptical about the atomic theory, disliking its apparent crudity and its poor fit with sciences outside chemistry. It was indeed embarrassing that Mendeleev's table classified sixty-three different elements, and therefore presumably different kinds of atom, and the number increased as predicted elements — and unpredicted ones — were found in subsequent years. Many of the elements were so similar that elaborate chemical and physical tests were necessary to distinguish them; and at a time when Darwin's theory of evolution was everywhere making headway, an evolutionary explanation of the periodic table was widely sought. In a return to a version of Prout's hypothesis, many chemists saw hydrogen as the progenitor of all the elements. They could not believe in a world having so many sorts of fundamental particle.

There is thus the irony that as chemists came to find that they could not do their science without the atomic theory, though there was no direct proof of it, they also found themselves, if they were reflective, profoundly unhappy with it. Some, like Kekulé himself, took refuge in the idea that the chemical atomic theory was not really a theory of matter at all. Lavoisier had urged chemists to avoid metaphysics and stick to the idea of elements operationally defined as the fundamental units of chemical reactions, rather than the basic consistuents of matter. His successors a century later left to physicists rather than metaphysicians the question of what matter was composed of, and saw chemistry as the science that dealt with the various kinds of matter rather than matter in general. Matter might be ultimately corpuscular or continuous; but all that the chemist needed to know was that atoms of elements were arranged in certain ways in compounds and crystals, and he did not need to concern himself with whether these atoms were really simple. In the nineteenth century, chemists became increasingly defensive towards practitioners of the new science of physics; while physicists tended to see them as a kind of glorified cooks. To chemists, it seemed a though their science was threatened with easy explanations in principle, worked out in armchairs, in place of detailed and painstaking laboratory investigations giving definite answers. They were even suspicious of physical methods, such as spectroscopy and polarimentry, which seemed alien to those used to crucibles, condensers and charcoal blocks. Most chemists were not expert mathematicians; blowing glass was more important than solving differential equations for the chemistry of the nineteenth century. To find genuine theories of matter in the latter part of that century we must turn to the physicists.

Further Reading

J.L. Heilbron, *Electricity in the 17th and 18th Centuries* (Berkeley, 1979)

D.M. Knight, *The Transcendental Part of Chemistry* (Folkestone, 1978)

D.M. Knight, *Ordering the World* (London, 1981)

D.M. Knight (ed.), *Classical Scientific Papers, Chemistry* (Mills and Boon, London, American Elsevier, New York, First Series 1968, Second Series 1970)

A.J. Rocke, *Chemical Atomism in the Nineteenth Century* (Ohio State U.P., Columbus, 1984)

C.A. Russell, *The History of Valency* (Leicester University Press, Leicester, 1971)

A. Wolf, *A History of Science, Technology and Philosophy in the 18th Century*, 2 vols. (2nd ed., London, 1962)

7 THE BIRTH OF MODERN PHYSICS

David Knight

Chemistry, as a practical art, has a very long history. But its transformation into a systematic science of matter, underpinned by a powerful explanatory theory based on plausible models and analogies, was brought about by the incorporation of a principle borrowed from physics, the principle of the conservation of mass. This principle had long been an axiom of atomism.

One can understand the separation of natural philosophy into physics and chemistry partly on the basis of their distinctive social roles and relations to different practical arts. Chemistry was related to metallurgy and medicine; physics to navigation, astronomy and gunnery. But a deeper distinction differentiates them. Physics came to be seen as the science of matter in general. Every moving body of whatever substance, must obey the laws of kinematics. Chemistry is the science of particular kinds of matter and aims to explain such phenomena as the interactions between acids and metals. But how were the underlying differences that are responsible for the radically distinct appearance and behaviour of organic and inorganic substances, of metals and non-metals, of medicinally active and inactive compounds and so on, to be explained? The quantitative version of the atomic theory provides the basis for convincing answers.

But at the end of the eighteenth century the exact territorial rights of these specialities had not been fully mapped out. This was partly due to the uncertain status of heat, light and electricity. Were they to be treated as part of chemistry as very attenuated material fluids (remember that some people thought the gases were compounds of elements with heat) or were they to be treated as phenomena as universal as the movements of any pieces of matter whatever? Many believed that the ultimate laws had been revealed by Newton. The method of mathematical analysis coupled with experiment that characterised physics is very ancient. It had already reached its modern form in Ancient Greece in the works of Archimedes. It flourished in the basic kinematic studies of the Merton College mathematicians of fourteenth-century Oxford.

Though the 'geographical centre' of physics had been well established two or more centuries before, its boundaries only became stable in the middle of the nineteenth-century, with the introduction of the principle of conservation of energy. Then heat, light, electricity and magnetism conceived as forms of energy could all be connected with mechanics into a new science which soon came to seem more fundamental than chemistry. Through the study of various kinds of radiation and of the nature of gases, nineteenth-century physicists arrived at theories of matter which were intended to be taken seriously as realistic models of ultimate reality. Physicists' theories were presented with varying degrees of assurance, but on the whole physicists seem to have been less ready than chemists to use a theory for purely practical or manipulative purposes. Explanation was their aim. As the eminent Cambridge man George Stokes (1819–1903) put it, a well-established theory should not merely link up observations but should serve to indicate the real processes by which nature brings about effects.

The Unity of Forces

Boundaries between sciences became frozen during the years soon after the full emergence of physics, no doubt largely because science began to be taught on a large scale at all levels, requiring agreed syllabuses, examiners and all the apparatus of a standardised curriculum. But in the early nineteenth century, the boundaries were fluid and polymaths abounded. By about 1800, philosophers in Germany — especially F.W.J. Schelling (1775–1854) — and men of science working in many fields began to place increasing stress upon the various forces of nature re-emphasising, against simple-minded materialists, the dynamic view of nature that had been held by such as Leibniz. Electricity and magnetism with their polar forces (positive and negative, north and south) seemed to underlie all sorts of changes where hitherto attention had been confined to surface effects. It was not the relatively inert and uninteresting matter which was important, but the forces which held it together and gave it its most interesting properties — substance is not what it weighs but where it acts.

Between 1780 and 1820, L. Galvani (1739–98), A. Volta (1745–1827) and then Davy showed how electricity and chemical affinity are manifestations of one power; William Herschel (1738–1822) found heat-rays beyond the red end of the solar spectrum, and J.W. Ritter (1776–1810) and W.H. Wollaston (1766–1828) found rays beyond the violet that were chemically active, thus bringing heat, light and chemical affinity into some kind of proximity. H.C. Oersted (1777–1851) after years of trying succeeded in demonstrating the connection of electricity and magnetism; and Faraday working in electro-chemistry and electromagnetism became increasingly convinced of the unity of forces, finally succeeding in rotating the plane of polarised light with a magnetic field to the astonishment of his contemporaries.

For those whose primary interest was in forces, the nature of the particles of matter was of less importance; and they tended to see them either as minute Newtonian corpuscles, or even as mere centres of forces, as they had been imagined by the eighteenth-century Jesuit Roger Boscovich (1711–87). His theory of matter had been devised as a kind of compromise between the views of Newton and of Leibniz. Leibniz had seen continuity everywhere in nature, and one of his objections to atomism was the formidable boundary between the massy hard particles and the void supposed to surround them. Boscovich's particles became mere points from which forces emanated.

Boscovich's atoms did not attract much attention except in Britain where the theory chimed in nicely with a native view of matter as a system of forces. The most notable convert was Joseph Priestley, the chemist and Unitarian theologian. He liked the theory as much because it seemed to make matter active, and thus suit his Christian materialism, as for its advantages in chemistry and electricity. Priestley's sympathies for the French Revolution and his unorthodox religion did not endear him to the next generation; and while Boscovich's atomic theory was discussed, especially in Scotland, in the early nineteenth century, it made few converts. It was often felt that on the one hand the theory only worked in general and not in detail — it could not be quantified or tested — and on the other that the world it depicted was not the real material world, which must be made of massy particles rather than of points.

Boscovich's theory was mentioned by Davy in a fragment of dialogue from a notebook published posthumously in 1840; but it was Davy's pupil Faraday who recalled public attention to it as a serious alternative to the hard and rigid corpuscles of the Newtonian atomic theory. Faraday had begun his studies of electrical conduction and induction with an idea of charges passing between contiguous particles; but he had shared Newton's distaste for 'action at a distance' — that is for matter acting where it was not. The working of gravitation had been a mystery for Newton and his contemporaries, but succeeding generations had come to accept gravity as a fundamental property of matter. Nevertheless to explain how the Earth and the Sun, or even two particles of gold in the 'solid' metal, attracted each other across void space was impossible, given the inert bounded atoms of the ordinary theory. Talk of hooks on atoms was begging the question, but so in a sense was talk of repulsive and attractive forces. Taken as an addendum to a solid atom theory it seemed to be only a more impressive way of redescribing the problem.

In 1842 in a lecture, supposed to have been given impromptu when C. Wheatstone (1802–75) the advertised speaker panicked and fled, Faraday outlined the view of matter that he had come to accept, and it was a Boscovichean one. The lecture was delivered at the Royal Institution. In the hands of Davy and Faraday the Institution had become a great centre for research in chemistry and physics, supported by lectures which became very fashionable. Eminent men of science, not necessarily connected full-time with the Institution, talked about their researches to a non-specialist audience; and sometimes the result was a less formal, less cautious paper than that usually read before a scientific society.

Faraday began with the problems of electrical conduction. Potassium metal is an excellent conductor. It is very light, and will contract on cooling, so that on the ordinary view of atoms there must be much empty space in a bar of potassium. It must therefore be the void which conducts; especially since potassium hydroxide, an insulator, actually contains more atoms of potassium in a given volume than the metal does. Similar reasoning about insulators such as sulphur leads to the paradoxical conclusion that the only continuous part of them must be the void — which must in this case be the insulator. Yet the void cannot have contradictory properties. The atomic model or analogue that saw the ultimate structure of matter as like an ensemble of billiard balls in empty space must therefore be inappropriate. Faraday argued it was best to assume as little as possible in one's model of the constitution of matter. He therefore proposed that atoms be thought of as points from which forces emanated, filling space. Where contemporaries saw a void, he saw networks of lines of force stretching out in all directions. Atoms no longer had boundaries, and it was the field between them that was where interest lay since it must be the ultimate reality. This idea was to transform our conception of matter. In one stroke physics returned to the field theory of William Gilbert and the early students of magnetism.

Faraday had this paper reprinted with his more formal ones of his *Electrical Researches*, which eventually ran to three volumes (1839–55). He republished another series of his papers as *Experimental Researches in Chemistry and Physics* (1859). The only paper reprinted in both series was his 'Thoughts on Ray-vibrations' of 1848: another Royal Institution lecture. He began with the remark that he 'merely threw out, as matter for speculation, the vague impressions of my mind'; but nobody publishes three times a paper in which they don't believe. This paper was an attempt to 'dismiss the aether, but not the vibrations'. In place of the aether envisaged as a medium for light waves and perhaps also for gravitation, he proposed that light and other forces were vibrations in lines of force. He declared that he did 'not perceive in any part of space, whether (to use the common phrase) vacant or filled with matter, anything but force and the lines in which they are exerted'. From a minimal Atomism, Faraday had gone over to a theory where matter was reduced to force. The 'forces of the atomic centres pervade (and make) all bodies, and also penetrate all space'. The mass of atoms, a primary quality for Atomists, was reduced to inertia, a force.

The essential step from Faraday's visionary glimpse of the true nature of matter to a fully fledged field theory was achieved by James Clark Maxwell (1831–79). By a dazzling use of dispensable mechanical analogies, his rotating tubes of fluid, he was able to develop a set of laws to describe the behaviour of the electromagnetic field which, if Faraday was right, was the ultimate stuff of all material substance whatever.

The Persistence of the Aether

Although his contemporaries recognised Faraday's genius, they were unable to follow his thought. Most of them were trained in applied mathematics, while Faraday was a chemist — he thought in pictures and they in equations. His view of a world of forces without aether or massy atoms — though worked out in its general outlines by philosophers of the previous century, such as Kant — was far ahead of its time. Probably because Faraday did not accept atoms of matter, he did not interpret his laws of electro-chemistry as indicating that electricity was itself atomic. This conclusion was drawn by H. Helmholtz (1821–94) in a lecture in memory of Faraday in 1881. It was a step on the way to the postulation on the electron with which, at the end of the century, a new kind of Atomism returned to physics. Helmholtz's reasoning was that if definite quantities of electricity deposit definite amounts of matter (as Faraday had shown), and matter is atomic, then electricity must also be atomic. Faraday would have denied the second premiss and so the conclusion. William Thomson (1824–1907), later created Lord Kelvin, was one of those responsible for enunciating the second law of thermodynamics and also for the successful Atlantic telegraph, and a contemporary of Maxwell. Unlike Faraday and Maxwell, but like many Frenchmen including J.B.J. Fourier (1759–1830), D.F. Arago (1786–1853), and A.M. Ampère (1775–1836), he saw the aether everywhere and tried to make mechanical models that were analogues of it. His vortex atom was proposed after he had come across a paper by Helmholtz on smoke rings; and it occurred to him that such rings in a frictionless fluid medium would be permanent and elastic. Matter would thus have been reduced not to force but to motion. New properties could not be ascribed *ad hoc* to vortex atoms, since there were strict mathematical rules governing smoke rings. They therefore escaped one of the major criticisms addressed for example at Daltonian Atomism, that properties of bodies were accounted for by giving the same properties to their component atoms — which really got one nowhere.

On the other hand, a frictionless fluid medium — the aether — was required for which there was no direct evidence whatever. If light were a wave motion, and many crucial experiments seemed to demonstrate that it was, then there must be something to undulate; if not, one would have a paradoxical situation, like Alice observing a grin without a cat. So the aether was inferred. The young J.J. Thomson's (1856–1940) first incursion into print in science in 1883 was with a prize essay developing the idea of the vortex atom, and demonstrating that up to six rings could be linked together in a stable configuration. Since this was the highest chemical valency then known, it really did seem that perhaps this model might fulfil the demands of physicists and chemists for an atomic theory. Despite such successes, the theory proved to be insufficiently fertile in predictions, and, even worse, inadequate to fit all the phenomena, and it remained a curiosity — an example of some elegant mathematics which turned out after all not to fit the world.

William Thomson's wrestlings with the aether continued and are particularly interestingly shown in the speculative lectures he delivered in the new Johns Hopkins University in Baltimore in 1884. This institution had been founded on the German model with emphasis on research as well as teaching, and his audience was composed of a small group of professors including A.A. Michelson and E.W. Morley, later of 'aether drift' fame. Thomson's lectures were not formally published until 1904. They showed him in great form trying out various models for the aether and expressing admiration for Boscovich's atomism. But he was not the confident classical physicist he sometimes appeared to be, and at the celebrations of his half-century as professor at Glasgow he declared that one word characterised his efforts, and that word was 'failure'.

This was not neurotic incapacity to recognise his own very genuine scientific achievements and no doubt was not completely serious but he did realise that many explanations — particularly those involving atoms and aether — were little better than verbal. His fear was always of falling into the trap of explaining the known and familiar in terms of the unfamiliar. With G.G. Stokes (1819–1903) he made the high demand of a theory that it should really be an attempt to represent reality and not just a calculus which saved the appearances. He made one of the first calculations of the upper limit of size for atoms — from the spreading of oil-films and other data — and was convinced of some kind of heterogeneity in nature; but he was unhappy about one of the most successful pieces of physics of his day: the dynamical or kinetic theory of gases. This was because assumptions about the equipartition of energy in molecules were built deep into the theory, and seemed to him implausible. On the other hand, the theory worked splendidly in terms of predictions, and had done much to unify chemistry and physics. Ultimately it provided direct evidence for the existence of atoms in the explanation of the Brownian Motion worked out in the opening years of the twentieth century. It is to this theory that we should now turn.

The Dynamical Theory of Gases

The theory that gases are made up of small particles in rapid movement, constantly colliding with each other and with the walls of a containing vessel, has long been of great interest to philosophers of science. Beginning with a model, a kind of three-dimensional game of billiards or snooker, men of science succeeded in improving its fit with nature and in using it to predict unexpected phenomena. The process fits the plan that William Whewell (1794–1866), the early Victorian polymath, had developed for the growth of many fields of knowledge. An original idea arrived at intuitively is refined and corrected by tests against experience while our conceptions of what is revealed in experience is influenced by our changing ideas.

The theory begins in the eighteenth century. Those in the previous century who had concerned themselves with the effects of pressure on gases and who believed some form of Atomism (like Boyle) seem to have held essentially static theories. For Boyle, air was springy because it was composed of springy particles like locks of wool or wood-shavings; while Newton thought particles exerted short-range repulsive forces on each other. Daniel Bernoulli (1700–1782), the most famous member of a dynasty of mathematicians, in 1738 propounded a kinetic theory of gases, but it was not significantly developed in the eighteenth century, although it featured in the world-view of G.L. Le Sage of Geneva who sought to explain gravity in terms of streams of fast-moving particles.

The great problem was one which struck Thomson: that the elasticity of gases was a much simpler phenomenon than that of solids, which was invoked to account for it. By the end of the eighteenth century, the 'air' of Boyle and Newton had been shown to consist of several gases, and other 'factitious' gases had been isolated or synthesised. All these gases were found to obey the same laws of expansion with rising temperature and contraction with rising pressure. Solids and liquids, by contrast, all behave differently from one another; so the physics of gases is much simpler than that of the other states of matter. By the early nineteenth century, gases were no longer thought to be as Lavoisier had conceived them — compounds including caloric — but as simple bodies lending themselves to mathematical treatment.

The difficulty was that they were generally supposed to be composed of atoms. But atoms were so hard as never to wear away or break in pieces, as Newton had put it. Elastic bodies like billiard balls are composed of many particles, and when they collide they become slightly deformed and then bounce back into shape as they separate. So elasticity can be explained for all composite bodies. We could even imagine perfect billiard balls which would not lose any energy in their collisions but would bound off each other endlessly. Really hard particles are another story altogether. In the seventeenth century Christopher Wren (1632–1723) had tried to work out dynamics of hard bodies and had found it very daunting. The great fire of London opened more fruitful possibilities to his genius, so he pursued the matter no further. Two unyielding bodies in head-on collision cannot bounce back and if the basic particles are really atoms, it is hard to see how their motion could continue. The static theories of gases — in which particles were not thought to be moving but merely vibrating or rotating — had the enormous advantage over a kinetic picture of moving particles that no collisions had to be supposed. However, as Dalton had found, the great problem was to account for diffusion, the mixing of gases, if one did not assume that the particles move rapidly.

In 1821 John Herapath (1790-1868), an unknown young man who later achieved some fame as a publisher of railway guides, put forward a kinetic theory in a paper he submitted to the Royal Society. Though full of interesting ideas the paper was ignored.

It was not until 1857 when Rudolf Clausius (1822–85) — like Thomson a pioneer of the second law of thermodynamics — took up the kinetic theory that became a central part of physics. Clausius worked out mean speeds of molecules, and in the following year he introduced the idea of a mean free path between collisions. In 1859 the theory was further improved by J. Clerk Maxwell (1831–79) later, as first Cavendish Professor at Cambridge, to preside over the construction of the Cavendish Laboratory. He was able to apply statistical ideas, using a model including the idea that the molecules were moving at various speeds. His work yielded the surprising prediction that the viscosity of a gas was independent of the density and varied with the absolute temperature. The theory now had the support of two of the greatest of nineteenth-century physicists. Moreover, it had a strong mathematical base, it had been used to make predictions, and it was closely involved with new ideas on temperature and energy.

By the 1870s, then, the theory was becoming generally accepted. Further, when two theories from different fields but based on the same idea begin to converge it seems that they must both be somewhere near the truth. This is what happens with the kinetic theory as the basic physics of gases and the molecular theories of chemistry came together in the opening years of the twentieth century. When W. Ramsay (1852–1916) was faced with the problem of determining the atomic weight of the new gases he had isolated — argon, neon, helium, krypton and xenon — he could not use the method of Cannizzaro (see p. 93) because these gases are so inert that no compounds of them could be found. Without atomic weights, they could not be placed in the periodic table. From their density, molecular weights could be determined, but the atomic weight depended upon whether there were one, two, or three atoms in their molecules. The kinetic theory depended upon the assumption that the energy of a molecule was equally partitioned between vibration, rotation and translation (or motion in a straight line). The simpler a molecule is, the less energy will be needed to make it vibrate and rotate, and therefore molecular complexity will affect specific heat. This can be quantified; and argon was found to fit the prediction for a monatomic gas. If it had only one atom in its molecule, then its atomic weight was 40, and it should be put between chlorine and potassium in the periodic table of the elements, and its congeners could be similarly placed. Here a theory which began as a model was being extrapolated into new territory. Important conclusions were being deduced from it which could not at the time be independently confirmed, but which fitted in well enough with chemical data to seem like a convergence.

A more direct link was with the work on the Brownian Motion of A. Einstein (1819–1955) in 1905 and of J.B. Perrin (1870–1942) in 1909. In 1828 Robert Brown (1773–1858), one of the greatest of botanists, wrote up observations he had made through the microscope on 'active molecules'. These were not formally published in his lifetime, appearing only in his *Works* of 1866; but he showed the phenomenon to visitors, in the natural history section of the British Museum where he worked, without explaining it. He found that grains of pollen in certain species of plants were 'filled with particles or granules of unusually large size, varying from nearly 1/4000th to about 1/5000 of an inch in length . . . While examining the form of these particles immersed in water, I observed many of them to be evidently in motion'. He called these particles 'molecules', in the general confusion of terms typical of the time, and he found that this 'seeming vitality' was evident in molecules coming from dried plants over a century old. Indeed, any organic matter, living or dead, bruised in water yielded up these particles; and so did fossils including coal and petrified wood, while the dust or soot rapidly deposited on everything in London was 'entirely composed of these molecules'. Even minerals from formations devoid of fossils, such as a fragment of the Sphinx, displayed activity. He wrote about 'spontaneous or inherent motion', and seems clearly to refer it to the particles rather than to the water.

Einstein and Perrin argued that the particles moved, not because they were active, but because they were bombarded by water molecules in their ceaseless motion. Einstein followed the motions of a single particle, while Perrin tried two approaches — counting the number of particles at two different depths in the liquid at equilibrium and observing the rate of diffusion in the liquid. The assumption they made was a bold one — that the kinetic theory could be applied to a liquid, and that the visible particles would have the same mean energy of agitation as water molecules and would diffuse with each other like a mixture of gases. Perrin's deductions agreed with those worked out from quite different evidence for actual molecular dimensions and convinced even the sceptical chemist Ostwald that matter really was atomic.

The Complex Atom

The Discovery of the Electron

The word 'atom' means something which cannot be split, and a complex atom is therefore really a contradiction. But the term atom had become attached to the simplest constituents of matter that preserved some chemical nature. Chemists were content to talk of atoms of carbon, hydrogen and so on in that sense, while admitting that they were probably made up of simpler particles. There were, in the nineteenth century, three lines of evidence for this. One was the analogies between families of elements as set out in the periodic table, with its evolutionary implications. The others were the phenomena of electric discharges through gases at low pressures and the study of the (very complex) spectra of the various elements. The convergence of these different lines meant that by the time Ostwald was converted to Atomism in 1911, the atoms which were postulated differed from those of Dalton in almost every possible way. This provides an excellent illustration of how scientific terms, like those of ordinary language, change their meaning over time in spite of efforts to define them more rigorously. If they are too rigorously defined they become useless, as theory is modified by experiment, observation, and reasoning. Flexibility of use and even perhaps long-term inconsistency in meaning are necessary for a science to be able to progress.

Davy in the 1820s was interested in the passage of electricity through gases, believing this would cast light on the nature of force and matter. The full investigation of the phenomena turned out to require better vacuum pumps than he had at his disposal. Right through the nineteenth century there was a slowly moving technical frontier, at any time limiting the possibilities of experiment. Improvements in pumps brought new discoveries, which in their turn modified theories. This is a relationship between science and technology which was important in other parts of science at this period. Chemistry depended upon improvements in apparatus — such as the introduction of platinum vessels and efficient condensers — and bacteriology depended upon the improvement of microscopes and thus ultimately, like spectroscopy, on the technique of glass-making.

Faraday attained lower pressures than Davy and observed the dark space around the cathode in a glowing discharge tube — a phenomenon which is now named after him. William Crookes (1832–1919) achieved still lower pressures and saw a further dark space, which as the pressure lowered came to fill the tube. A new radiation, the cathode rays, then made itself evident. Crookes was not a professional scientist, as Davy and Farady can perhaps be described. There are some fascinating remarks about him made by J.J. Thomson, his successor in cathode ray work, in his autobiography. Crookes, apparently, had waxed moustaches, and looked rather foreign; he had at one time owned a gold-mine. He was a good scientific journalist, running the gossipy and speculative *Chemical News* which proved an excellent complement to the arid *Journal of the Chemical Society* and provided a model for the general weekly journal, *Nature*. Crookes made a name for himself with his discovery of the element thallium. He invented tinted lenses by including salts of the Rare Earth elements, like lanthanum, in the glass. He took on a good deal of consultancy and analytical work. Later he became greatly involved in spiritualism, the boundary between the material and the immaterial always fascinating him. He was one of the great men of Victorian science. To Thomson he seemed to approach the cathode rays like an explorer in a new country noting down anything and everything he saw. This is not really a fair characterisation of Crookes' work, but like a good caricature it has something in it. Crookes showed that the cathode rays come in straight lines from the cathode, are deflected by a magnetic field, will make a small paddle-wheel rotate, cause zinc sulphide to fluoresce and cast sharp shadows of, for example, a Maltese cross. He saw the rays as the effect of matter in a form so tenuous that it was in a fourth state, even simpler than the gaseous state, where the free paths of the molecules were comparable to the dimensions of the tube. For him, the rays were composed of particles, but not, he thought, of sub-atomic size.

German investigators, notably H.R. Hertz (1857–94) the discoverer of radio waves, interpreted the cathode rays as a wave motion. In favour of their view was the fact, so it seemed, that the rays could not be deflected by an electric field. J.J. Thomson took up the study of cathode rays after his work on vortices. In 1895 W.K. Rontgen (1845–1923) discovered X-rays — emanations capable of blackening photographic plates — coming from a cathode ray tube and took the first famous X-ray photograph of his own hand. The amazing properties of X-rays increased interest in the cathode rays which generated them. In 1897 Thomson succeeded in showing that Hertz and others had failed to deflect the rays by an electric field because they had not achieved good enough vacua in their tubes. In his apparatus he deflected the rays with a magnetic field, and then applied an electric field to bring them back again to where they fell undeflected. This enabled him to calculate the ratio of mass to charge of the particles of which the rays were composed. He found that they were extremely small, or else enormously charged, their ratio of charge to mass being over a thousand times greater than that of a hydrogen ion. Further experiments by Thomson confirmed that the charge was of the same order as that of the hydrogen ion and the particle much smaller. These results were refined by R.A. Millikan (1868–1953) in his famous oil-drop experiments of 1909–12.

Thomson called the particles composing the rays 'corpuscles', consciously reviving the theory of his seventeenth-century predecessors. The rays had been shown to be the same whatever the cathode was made of and whatever gas was in the tube; so it was plausible to conclude that they were the real atoms, the ultimate constituents of matter, postulated by Democritus, Boyle and Newton. The problem was that they had an electric charge, while ordinary matter is neutral. Thomson first identified them with the electrons proposed as atoms of electricity by G.J. Stoney (1826–1911) independently of, and indeed before, Helmholtz. He then proposed an atomic model in which the electrons were immersed in positive material, like plums in a plum-pudding, in such a way that they would be as far apart from each other as possible. This plum-pudding atom was the model generally adopted in the opening years of the twentieth century.

The Explanation of the Spectrum

The other line of investigation was through spectra. It had been long known as a spot-test in analysis (and exploited in firework-making) that various metals gave a characteristic colour to flames; but one problem about using this as a test was the ubiquity of sodium salts in days when standards of chemical purity were not as high as they are nowadays. Almost everything would give a flame the brilliant orange colour of the sodium flame, which might well mask other components. It was Josef Fraunhofer (1787–1826), in the second and third decades of the nineteenth century, who succeeded in controlling the making of optical glass with such precision that dark lines characteristic of the continuous solar spectrum could be identified. In 1860 R.W. Bunsen (1811–99) and G.R. Kirchoff (1824–87) at Heidelberg inaugurated chemical spectroscopy, demonstrating that on heating each element emits light with a characteristic spectrum of bright lines; when, as in the Sun, superimposed on a continuous spectrum from a hotter source, these same lines appear dark. They announced in rapid succession the discovery of two new alkali metals, caesium and rubidium, which they had identified from their spectra. This method of analysis, the first to depend on 'physical' methods rather than on 'wet' or 'dry' chemical tests, soon became standard, and was, for example, employed by Crookes in the discovery of thallium.

The surprising thing was the complexity of the spectra of the elements. If their atoms were like billiard balls, then they could be made to vibrate, and their vibrations at certain appropriate frequences could be supposed to yield certain lines in the spectrum. But billiard balls can vibrate only in relatively simple ways, while there are many lines for each element. Even hydrogen, which one might with Prout consider the only real element, had so many lines that it was hard to believe it was not complex. There appeared to be no rhyme nor reason behind the various lines characteristic of each element. All that could be done was to try to map them. Often this proved very difficult. Crookes was unable to separate the Rare Earth elements with techniques then available, so that each seemed to merge into the next. He concluded that they were mere varieties, or meta-elements, rather than true species.

In 1884 J.J.Balmer (1825–98), a Swiss schoolmaster of nearly sixty, announced that he had discovered the mathematical relationship of the lines of the hydrogen spectrum: the frequencies for a converging series, now called the Balmer series. Similar series were later found to fit other parts of the hydrogen spectrum beyond the visible light region; and Balmer sought regularities in the spectra of helium and lithium. At the time, these numerical investigations seemed more ingenious than important, rather like the search for numerical ratios among atomic weights which had guided some chemists towards the periodic table; and it was as qualitative evidence of complexity rather than as quantitative evidence for specific structural hypotheses that spectra were of importance in the late nineteenth century.

In 1898 Ernest Rutherford (1871–1937), who had come from New Zealand to Cambridge and worked with J.J. Thomson, began his work on radio-activity, moving in that year to a post at McGill University in Canada. This phenomenon had been discovered in France by Henri Becquerel (1852–1908), one of a line of distinguished scientists; while Marie Curie (1867–1934) had isolated two active elements from pitchblende — radium and polonium. She was a chemist and looked for chemical substances producing the radiation. Rutherford was a physicist and sought the cause of the phenomenon. He came to the conclusion that there was what he called a sub-atomic chemical change going on, in which one element was actually turning into another. This new alchemy would have been deemed impossible by a strict Daltonian, but it caused little surprise among men of science who had been prepared for such things by the evolutionary speculations of Crookes and others. Rutherford's collaborator, Frederick Soddy (1877–1956), came to London to work with Ramsay and they actually followed a transmutation in the laboratory, as a sample of radon gas after a time began to show a helium spectrum.

The Nuclear Atom

In 1907 Rutherford moved to Manchester, where he found that when thin sheets of gold or platinum were bombarded with alpha rays (which he had already shown to be helium atoms with a positive charge) most went straight through, but about one in eight thousand was deflected through more than ninety degrees. This was a very surprising find, for if the atoms were like plum-puddings with negative electrons scattered through a kind of positively charged matrix then every projectile passing through would be slowed up and perhaps somewhat deflected but one would not find this occasional big deflection. Rutherford's conclusion was that atoms were quite different in their internal architecture; the experiment could be explained if they had a nucleus where the mass and the positive charge was concentrated, surrounded by electrons in orbits. Most of the atom was empty space, through which alpha particles could easily pass; but one getting near a nucleus would be subject to enormous forces and diverted through a large angle. Like those of Boscovich, these new atoms undermined the whole idea of solidity. They occupied territory more like soldiers with rifles than tennis-balls filling a box.

The great problem about this atomic model was that it was unstable. A charged particle moving around an oppositely-charged nucleus should rapidly spiral into it, emitting energy; whereas atoms have to be stable indefinitely. The solution to this problem depended upon quantum theory, devised to account for radiation from black bodies and upon Balmer's series for the spectral lines of the hydrogen atom.

Something which radiates light, heat, etc. of all possible wavelengths is called a 'black body'. Black-body radiation was a problem in the late nineteenth century because it did not conform to the predictions made from the wave theory of light. This theory since its revival by Thomas Young and A.J. Fresnel in the first quarter of the century had survived crucial experiments with triumphant success, so that by the 1870s it had come to be accepted as a fact. John Tyndall (1820–93), the eminent physicist, urged chemists to accept an atomic theory although it was strictly unprovable, just as everybody accepted light waves. But no tinkering with the theory made it possible to fit it properly to black bodies. Between 1897 and 1901 Max Planck tried the idea that energy, like matter, came in lumps, or quanta. This saved the phenomena in that, with this assumption, Planck was able to explain the black-body spectrum. But he was a reluctant hero, unwilling to see the revolutionary implications of what he had done. It was Einstein, in treating the photoelectric effect in 1905, who showed that light is quantised, coming in photons — a concept which seemed alarmingly like that of the light particles of the eighteenth century. One of the best established of scientific theories — that light was the effect of a wave-motion in a continuous medium — had been shown to be untrue, or at least applicable only within certain limits.

The application of quantum theory to atoms was made by Niels Bohr (1885–1962). He came from Denmark to Cambridge in 1911 to work with J.J. Thomson. He found Thomson very busy, and allowed Rutherford (whose nuclear atomic model had recently been published) to entice him away to Manchester. Bohr made the daring assumption that the orbits of the electrons are quantised, and that the spectral lines corresponded to jumps from one permitted orbit to another. This assumption permitted him to connect the structure of the hydrogen atom and the structure of its spectrum quantitatively, so that the new atomic model began to seem much less *ad hoc*.

Rutherford had been able to compute the positive charge on the nucleus of gold from his scattering experiments. Meanwhile, in the hands of Max von Laue (1879–1960) and then of William (1862–1942) and Lawrence Bragg (1893–1971), X-ray analysis was being applied to crystals; while Henry Moseley (1887–1915), working with Rutherford, found that part of the X-ray spectra of elements (which were much simpler than the visible spectra), the K-lines, changed by a regular increment as one went up the periodic table. One could now at last decide where elements remained to be discovered, particularly among the Rare Earths which had so perplexed Crookes. An element could be defined in terms of its atomic number — that is the positive electric charge on it nucleus, rather than its atomic weight which reflected the total mass of its constituents. The periodic table had been reduced to a piece of physics, depending on orbits of electrons arranged according to quantum numbers around a central nucleus, the nucleus of atoms of each element having a particular charge. This reduction made surprisingly little difference to chemists. They still used the table qualitatively, rather than working from fundamental mathematical laws involving quantum numbers for making chemical predictions. Indeed, for everything except hydrogen the difficulties of computing chemical properties on the basis of the physics of the atomic architecture remained insuperable. It was from chemical properties that one still had to work out how the electrons go, and not the other way round.

The atom of Rutherford and Bohr thus represented a remarkable simplification compared to the Daltonian model. Instead of nearly ninety irreducibly different particles representing all the chemical elements, it required only two: the negatively charged electron and the positively charged proton. According to Rutherford and Bohr the nucleus was made up of protons, with some electrons to balance the nett charge with the numbers of proton required to have the appropriate mass; and around it went the rest of the electrons in quantised orbits. For Bohr, these had been circular. Arnold Sommerfeld (1868–1951) suggested in 1915 that they should be elliptical, their ellipticity being governed by another quantum number — which was confirmed by studies of the fine structure of spectra. Gradually since then models of atoms have become more complex again, and the number of fundamental particles has once again steadily grown. The idea of the electron as a particle had apparently been confirmed in J.J. Thomson's famous experiment. The logic of that experiment required the assumption that the electron must be either a particle or a wave. This was the basic assumption of all classical theories of matter, but, to the consternation of most physicists, it soon had to be given up. The nature of matter, its relationship to energy, and the mode of being of the truly fundamental particles that make up the nucleus — these were the questions to agitate anybody concerned with the theory of matter in the half-century after Rutherford's model appeared.

Further Reading

W.H. Brock, *From Protyle to Proton: William Prout and the Nature of Matter, 1785–1985* (Adam Hilger, Bristol, 1985)
D. Gooding and F.A.J.L. James (eds) *Faraday Rediscovered* (Macmillan, London, 1985)
M.E. Hesse *Forces and Fields* (Nelson, London, 1961)
T.H. Levere *Affinity and Matter* (Oxford University Press, Oxford, 1971)
B. Schonland *The Atomists* (Clarendon Press, Oxford, 1968)
S. Wright (ed.) *Classical Scientific Papers: Physics* (Mills and Boon, London, 1964)

8 PHYSICS AND CHEMISTRY IN THE MODERN ERA

David Knight

The Ambiguity of the Nature of Matter

Einstein's work on the photo-electric effect had shown that one could no longer simply say that light was a wave motion, nor indeed that it was composed of an unorganised stream of corpuscles, for the classic experiments of the previous century on diffraction and on the velocity of light in air and in water had falsified that. In still seemed that Thomson had shown that the electron was a particle. But it occurred to Louis de Broglie (1892–1960) that if light had two aspects, so might matter; in particular electrons might behave as both waves and particles. Not long after the First World War wave-like behaviour of a material body was duly observed. This discovery — electron diffraction — made the electron microscope possible, with its much higher resolving power than can be obtained with visible light. To imagine an electron as a particle was therefore very partial, and in the wave mechanics of Erwin Schrodinger (1887–1961) of 1926 an attempt was made to interpret matter as waves. But the general view became that of Bohr, who in 1927 put forward his principle of complementarity.

According to this idea, sub-atomic entities display wave and particle aspects, each in appropriate circumstances, and the two theories between them cover all possible ways the physical world can be observed. On the other hand, because the apparatus to detect these entities must interact with them (and so in a sense must the observer), an element of uncertainty is introduced: the more accurately one measures a 'particle' property such as position, the less accurately can one determine a 'wave' property like momentum, and vice versa. The claim of P.S. Laplace at the beginning of our period — that anybody knowing the position and velocity of every particle in the universe would know the past and the future — becomes empty: nobody could in principle know these things.

This may be seen as a threat to determinism and causality, and indeed for 'complementarity' some people read 'indeterminacy'. Threats to causal principles go back further — perhaps to Maxwell's work on the kinetic theory of gases and his demonstration that the second law of thermodynamics rested on statistical grounds, and certainly to Rutherford's experiments in the first decade of the twentieth century, in which he found that radio-active decay while completely regular in large numbers of atoms seemed not to be caused in particular cases. For any particular atom, the amount of time it spent before decaying seemed to be a matter of chance. The great Victorian physicist and astronomer Sir John Herschel (1792–1871) had written of Darwin's theory as the 'law of higgledy-piggledy', but his own science had by the 1920s come to rest on much the same foundation. Darwinians argued that it was not causality which was essential to science, but the applicability of laws even if they held only for large aggregates of individuals. Physicists had to argue the same way, taking consolation from the way larger-scale processes do happen regularly.

The Identity of Mass and Energy

The distinction between matter and energy, upon which the science of physics had been founded, also became blurred in the early twentieth century. Quantum theory had energy coming in a kind of atomic form, while electron diffraction had matter behaving as a form of energy does, since waves are undulations, not streams of particles. Einstein formulated his famous equation for conservation of mass-energy $E=Mc^2$ (where c is the velocity of light), which replaced the two more primitive laws according to which each was conserved separately. This equation implies that matter can be converted into energy, and vice versa, and it accounted for the great quantities of energy associated with radioactive changes. The series of sub-atomic chemical changes which Rutherford and others followed led to products which had less mass than the starting materials — making them quite different from the large-scale chemical changes which Lavoisier had investigated in the work which had led to the overthrow of the phlogiston theory. The mass lost had turned into energy, and because the velocity of light is so enormous, a little mass yields a great deal of energy.

Assuming that this conversion process was going on in the core of the Earth enabled geologists to give our planet the long history they wanted for the slow evolutionary processes required by Darwinian theory. Previously the Earth had been supposed to be a great mass slowly cooling, and by extrapolation backwards one could demonstrate (as William Thomson did) that longer ago than fifty to one hundred million years it would have been too hot for any kind of life. If on the other hand the Earth's stock of uranium and thorium was gradually decaying with emission of energy, then the Earth's temperature could have been roughly constant for very much longer. Here T.H. Huxley's (1825–95) biological argument turned out to have been right, and physicists who had confidently appealed to the second law of thermodynamics wrong. This episode led to some loss of confidence in arguments from physics, and geologists were reluctant to bother with physical evidence too much in the next half-century.

Thomson had also computed the age of the Sun, making all the reasonable assumptions available — such as, that it was made of coal, that it was also fuelled by meteors crashing into it, and that it was collapsing on itself under gravity and thus emitting energy. All these assumptions together did not give it a very long life in geological terms. Its age was, Thomson concluded, about the same as that he had assigned to the Earth. The spectroscope was used in the 1860s to determine what the Sun and the stars consisted of, and the Sun was found to be largely hydrogen, with also a new element unknown then on Earth and christened 'helium'. With Einstein's theory of mass-energy, the Sun could be seen not to depend on Thomson's processes for its heat, but on the conversion of hydrogen to helium — what we would call a thermonuclear process — with the loss in mass going into energy.

T.H. Huxley could write in the nineteenth century that unlike organised religion, science did nobody any harm: gunpowder after all had been invented by a friar, while science had given fertilisers, dyes, railways, telegraphs and other generally desirable things. In the nineteenth century the man in the street did not need to concern himself much with theories of matter — though there were in fact large audiences for scientific lectures, and a market for books that were popular but pretty stiff, among working men as well as the more fashionable groups which went to the Royal Institution. The nature of the atom only came to concern us all after 1939, when the implications of Einstein's equation for weapons became clear, and were demonstrated in the atomic bombs of 1945. Since then we have been living in the shadow of nuclear war, so that the optimism of nineteenth-century scientism sounds very hollow, and the scientific debates about the existence of atoms innocent and ivory-towered.

Energy can also on Einstein's view be converted into mass, and this happens for example as particles are accelerated towards the velocity of light. This represents the speed-limit of the universe in the new physics, since as they come closer and closer to this velocity particles increase rapidly in mass — an increase which seems to tend to infinity. This was found to be experimentally true of electrons. This introduces the distinction between some fundamental rest-mass, and effective mass in motion. Particles proposed in modern physics may have no rest-mass, but nevertheless do have effective mass by virtue of their motion.

Fundamental Particles

The Constituents of Atoms

This brings us to the third aspect of theory of matter in the first half of the twentieth century: the nature of the fundamental particles. The happy days of Rutherford and Bohr's atomic model with its two kinds of particle lasted through the 1920s. It was refined by the incorporation of elliptical orbits. Wolfgang Pauli in 1924 suggested that no more than two electrons could occupy each orbit. If there are two, they must differ somehow. This difference was located in a property that was called 'spin'. This idea was called the exclusion principle, and allowed much existing knowledge of atoms and spectra to fall into place.

It was still supposed that the nucleus consisted mainly of protons with rather less electrons to balance out the total charge. But in 1932 James Chadwick (b. 1891), in the Cavendish Laboratory where he was working with Rutherford, interpreted experiments on the radiation from beryllium excited by alpha-rays as due to a new kind of particle: the neutron. This was of the same mass as the proton but was electrically neutral. The nucleus was henceforward seen as made up of protons and neutrons. The total mass of a nucleus was due to the protons and neutrons, its charge to the protons alone. There was no need to suppose that the nucleus contained any electrons. This solved various theoretical problems and indicated a valuable way of splitting atoms since neutrons are not repelled from muclei as positively charged particles are, but it did begin to make atomic models more complex again.

Another neutral particle was proposed by Pauli under the name 'neutrino'. This was more controversial than the neutron, because it was inferred by a theoretical physicist rather than in any sense observed. In radioactivity involving emission of beta rays (or electrons) — called beta-decay — the electrons emitted formed a continuous spectrum whereas one would have expected them to have discrete energies. Moreover energy and mass did not seem to be conserved during the process. To save the conservation rules, Pauli proposed very small neutral particles, undetectable by any means then available and having spin like the electron. It was not until the end of our period, many years after Pauli's proposal, that anything like direct evidence for the existence of neutrinos was forthcoming. The acceptance of a particle as remote from the things of common experience as the neutrino shows how untrammelled is creative thinking in physics. It is not a matter of mere refined commonsense.

Electrons have a negative charge; and in the formal theory of Paul Dirac (b. 1902) of 1928 there was room for a particle of the same size but opposite charge, though contemporaries did not recognise this. In 1932 Carl D. Anderson (b. 1905) in California identified such a particle, though he knew nothing of Dirac's theory, and called it the positron. This was the first of the class of anti-particles. Whenever a particle and its anti-particle — such as an electron and a positron — meet they annihilate each other with emission of energy. Like the electron, the positron involved a curious and complicated convergence of theory and observation rather different from the old picture of the testing of a theory by reference to its observable predictions and of the generalising from facts — schemata which were often supposed to encapsulate scientific method. Anderson's discovery led eventually to Dirac's theory being taken more seriously, which in its turn led to the discovery being reinterpreted in the light of the theory.

In the nineteenth century physics was generally a cheap science. Faraday reckoned that if he ever needed a big piece of apparatus, he would have no trouble in getting the backers of the Royal Institution to pay up for it. In the middle of the nineteenth century the Royal Society had an annual grant from the government of £1,000 to back research projects, and the British Association also had some funds at its disposal, chiefly from subscriptions. In Germany and the USA rather more was available, but on the whole it was astronomy, life sciences and earth sciences which were expensive. To send out a survey ship like HMS *Beagle* or HMS *Challenger* cost a lot of money and so did the upkeep and equipment of observatories for astronomy, seismology and meteorology. Much physics down to the 1920s was done with cheap apparatus, if not strictly with the legendary sealing-wax and string.

Atomic and nuclear physics began to demand expensive apparatus, staffed by teams, as observatories had been in an earlier period. The days when a leading physics laboratory could be worked by one man like Faraday, with perhaps a technician or assistant to help out, have largely gone. Physicists began to separate into theoreticians like Dirac and Pauli, and experimentalists like Anderson; and the experimentalists devised increasingly large and costly pieces of apparatus to supply and concentrate the energy needed to produce the particles and interactions they wanted to study. One cheap source of high-energy particles was cosmic rays, and various fundamental particles have been detected on photographic plates exposed to cosmic radiation. But cosmic rays cannot be produced to order. The fact that charged particles can be accelerated and directed by electric and magnetic fields has been exploited in the constructions of new but costly 'accelerators' which now provide energetic particles for most modern research into sub-atomic reactions.

By 1960, then, physicists had developed an atomic model, consisting of a complex but compact nucleus with a haze of orbital electrons. At this time the model began to be extended into chemistry, as molecular orbital theory, to make obsolete the ball and wire models of molecular structure introduced a century before. Atoms were known to be composed of protons, neutrons and electrons; but there was a series of other particles, like neutrinos and mesons, some of them of very short life, the role of which was unclear. But despite deep remaining puzzles, it was obvious that much more was known about matter than ever before. The new atomic models of the 1920s and '30s had meant for example that the solid state was now much better understood than it had been. Indeed the conduction of electricity through gases, which had been so much studied in the late nineteenth century, was now more problematic than that in solids. The fruit of this new understanding was the transistor, which came to replace the gas-filled diodes and triodes as the basis of electronic equipment in the 1950s. As Faraday had hoped, the conduction of electricity, interpreted as the flow of electrons, had indeed cast light on the nature of matter.

Unlike Faraday's atomic model taken from Boscovich and confessedly minimal, the nuclear atom could provide a number of quantitative predictions and explanations. The old problems of explaining elasticity and hardness were no longer those which perplexed men of science, but the old danger of trying to explain the familiar in terms of the recondite remained. What is perhaps most striking about the story is that the word 'atom' came to change its meaning so completely during these two centuries. The hard, massy and probably spherical particles which were different for each element gave way to complex arrangements of variously-charged smaller particles. Just as the apparatus of science and the divisions between the sciences change, so the concepts used are refined in contact with experience so that over a period they may almost reverse their meaning. Yet while in the sciences as in other things there is change, there is also continuity, and the programme of the Atomists of Antiquity, of accounting for all the complicated and various things we see in the world in terms of atoms having much simpler properties, remains the aim of physicists to this day. But recently events have taken a disconcerting turn.

The Proliferation of 'Particles'

By 1960 about a hundred fundamental particles had been recognised, some of them exceedingly unstable and some seeming much more fundamental than others. Once again, some sort of system of classification was pressingly necessary, as it had seemed to Mendeleev a hundred years before as he contemplated the variety of chemical elements. We are apt to think of classification as a sort of 'natural history stage' through which all sciences pass in their youth before they grow into something handsomer, more mathematical and explanatory. But in fact we classify all the time, and classification is a highly theory-laden activity. Physics depends upon ordering its fundamental entities just as much today as it did in the days when these were thought to be Earth, Air, Fire and Water. What one thinks one is classifying may make a big difference to the system of classificatory categories one chooses.

In contemporary physics, there are those of Faraday's cast of mind who turn their attention away from the particles to the relations between them. Looking for increasingly fundamental particles seems to let us in for a regress of indefinite extent. We say that the air is not an element, but is composed of particles of oxygen, nitrogen and so on; that these particles are really molecules, composed of a number of atoms; and that these atoms are composed of more fundamental components. At each stage in this progress it looked as though an end in something really fundamental had been reached; but the next generation pushed through it to some smaller and more fundamental particle. At each stage the properties of the particles became more remote from those of ordinary things, and the risk seen by Lord Kelvin of explaining the familiar in terms of the unfamiliar and the simpler in terms of the more complicated became greater. The sub-atomic particles required for quantum physics are so odd when compared to rocks and cannon balls that it may be that they are like phlogiston: theoretical entities postulated or inferred rather than discovered and destined to be abandoned when a more coherent way of looking at the phenomena of cloud-chambers, cosmic-ray photographs and accelerators is proposed.

Thus the physics described by Gary Zukav in his very entertaining book, *The Dancing Wu-Li Masters*, is that of a cosmic dance. It is not the particles, whose continuous existence is probably illusory, but their patterns that should be studied, just as in a complicated folk-dance it is the movement rather than the individual dancer that one follows. Matter is the momentary manifestation of interacting fields. We have to cast off ordinary thought-processes, characteristic even of Galileo, Newton and Lavoisier, and turn instead to mystics of the East and West who have managed to think without allowing themselves to be intimidated by apparent contradictions and to see space, time and matter very differently from more mundane folk wrestling with a railway timetable or the packing of a suitcase. This opens up exciting if rather vague hopes of physics becoming once again a study which opens the mind to cosmic contemplation, as it has been for Newton and Boyle. But while everybody would concede that Laplace's vision of 1800, of a being who knew the position and velocity of every particle in the universe and therefore saw the complete past and the future, must be abandoned, the cosmic dance is still perhaps a little too 'Californian' a conception to prevail easily everywhere.

In more traditional vein, as by 1960 the number of particles increased, Murray Gell-Mann (b. 1929) and Yuval Ne'eman (b. 1925) proposed a new classification called The Eight-Fold Way. Despite its Eastern overtones, this was really a kind of periodic table. Just as Mendeleev's had justified itself by predicting unknown elements, so this classification led to the discovery of the Omega-minus particle, which is extremely rare and short-lived. Just as Mendeleev's table inspired many — but not at first Mendeleev — to offer evolutionary explanations of the variety of elements, so this system seemed to cry out for some kind of explanation. If one's system is really natural, reflecting the way things are and not some artificial or merely practical device like dividing animals into 'game' and 'vermin', then there ought to be some reason behind it.

In 1964 Gell-Mann and George Zweig independently proposed that there were more fundamental units than any so far discovered, which came to be called 'quarks' — borrowing a word from James Joyce's *Finnegan's Wake*. All known particles are supposed to be composed of a few different kinds of quark. Amongst their more curious properties, they have ⅓ unit of electric charge. Protons are thus composed of three of them. Nobody has yet seen or detected a free quark in any way and it may be that to isolate a single one is impossible. The hope is that they are so fundamental that, if one were to be found, the question 'what is is made of?' would be misconceived. It seems to be a law of nature, in some very fundamental sense that quarks are always confined within the protons, mesons and so on which they compose.

A difficulty of this model or picture of the ultimate constitution of matter is that there seem to be many differnt sorts of quarks, which are described as having various 'colours' and 'flavours', and as being 'up' or 'down' — quite arbitrary ways of describing their differences. A new periodic table seems to be needed once again. It seems to be one thing to classify easily identifiable materials like iron, gold and chlorine in a table and quite another to do it for particles which are not merely undetected but perhaps undetectable. There is a story of a Scottish professor who complained about this very troublesome fellow, Humphry Davy: 'Just as we had got the chemistry syllabus straight, he discovered a new element, and the whole thing had to be reorganised.' This is a process happening continually in modern physics: just as a theory seems to embrace everything, some new phenomenon is discovered and the apple-cart is upset again.

Many of the particles on the present list have very short lives, that of the Omega-minus particle being one billionth (10^{-9}) seconds; others exist for much longer, like neutrons (918 seconds); and others again seem, like protons and electrons, to have an infinite or anyway indefinite existence. The atoms of Lucretius were everlasting, as were those of Newton, for they could never wear nor break. Rutherford's atoms could break in pieces, and this marked a major departure from the whole atomic tradition. Relativistic physics allowed the transformation of mass into energy, and therefore it became possible to conceive of the disappearance and reappearance of particles. The old programme of explaining changes in terms of the arrangements — couplings and uncouplings of unchanging atoms — has ceased to be possible in principle because the fundamenal particles are themselves changing. What they do retain of the old properties is identity. All neutrons, all electrons and so on are exactly alike.

But the particles that seem to us to have indefinitely long lives may not last for ever, though they do not gradually wear out like we do. It is possible that protons for example have a limited life and that they are slowly disappearing. What the physics of the twentieth century has done is to give us matter which had a beginning in the 'big bang', a vast explosion, from which our cosmos originated — matter which can be transformed into energy and which may have only a limited span of existence. This is very different from the atoms of Newton and Dalton, the unchanging entities subsisting through all the processes going on in the world. To find such unchanging entities now looks very difficult if not impossible. Physics has become a much more dynamic science, a study of the flux of more Heraclitean things. We live in an exciting time, when there seems to be little that is certain about the nature of matter. The connections that Galileo and Newton made between the physics of the laboratory and that of the heavens have proved to be extremely fruitful, so that studies of the spectrum of the Sun, stars and nebulae in the nineteenth century, and of cosmic rays, radio stars and 'black holes' in the twentieth, have cast much light on the nature of matter and its changes. History does not repeat itself, but physicists are working in a long tradition which cannot but guide them whether or not they are aware of it. Parts of it will survive and parts will be abandoned; but whether with the theory of quarks we have really got to the bottom of things, time alone will show.

Further Reading

D. Bohm, *Causality and Chance in Modern Physics* (Routledge and Kegan Paul, London, 1957)
G. Hartcup and T.E. Allibone, *Cockcroft and the Atom* (Adam Hilger, Bristol, 1984)
J. Hendry, *Cambridge Physics in the Thirties* (Adam Hilger, Bristol, 1984)
Sir G.S. Thomson, 'Matter and Radiation' in R. Harré (ed.) *Scientific Thought (1900–1960)* (Clarendon Press, Oxford, 1969)
J.S. Trefil, *From Atoms to Quarks* (Scribner, New York, 1980)
G. Zukav, *The Dancing Wu-Li Masters* (Bantam, New York, 1979)

9 SOCIAL INFLUENCES IN MATTER-THEORY

David Bloor

Introduction

As historians press their studies of the growth of science into greater detail, one fact emerges with ever increasing clarity: this is the contribution that scientists themselves make to the end-product that we call 'scientific knowledge'. Scientists are not passive recipients of truths furnished by experience so much as active partners in a process of construction. No one has ever doubted the importance to science of the observable effects provided by the experimental arrangements that scientists contrive; but the ways that scientists interpret their findings and attach meanings to them are equally important. Acts of theoretical interpretations are a necessary part of science, because the raw data of experiment cannot by itself uniquely determine our understanding. This well known feature of knowledge is called 'the underdetermination of theory by fact'. Philosophers (who were among the first to emphasise the importance of 'underdetermination') want to know the best, or most rational, method of closing the gap between data and theory. Should scientists be reluctant to go beyond their data, or should they take risks and speculate boldly? Historians, whose professional concerns are more descriptive than prescriptive, will want to know exactly how scientists in fact reach their conclusions, rather than how they ought to proceed. Sociologists share this matter-of-fact curiosity and are particularly interested in the role played by tradition and inherited culture, by the detailed structure of the scientific profession and its relation to other institutions, by the commitment that a group has to a particular method of approach or form of scientific instrumentation and even by the degree to which scientists feel confident and valued rather than threatened and beleaguered. It is not always easy to see how factors such as these might be at work in the process of constructing scientific knowledge. The aim of this chapter will be to provide some simple illustrations of precisely how social processes are inextricably bound up with the history of matter-theory. The examples will be drawn from the writings of a number of different historians. By way of introduction to this material, it may be useful to illustrate the fact of underdetermination itself, to show how it helps raise questions of a recognisably sociological kind.

Underdetermination and Common Conceptions

In his fascinating book *The Tiger and the Shark* the historian Bruce Wheaton describes the experiment which provided physicists with the first plausible measure of how rapidly X-rays moved.[1] Although X-rays had been discovered in 1895, it was ten years before Erich Marx (1874–1956) devised an apparatus that seemed to permit a measure of their speed. A diagram of his apparatus, taken from the *Philosophical Magazine* of 1907, is given in the figure below. Marx's apparatus consisted of two vacuum tubes. One of them, an X-ray tube, enclosed the electrodes C and A; the other enclosed the electrodes marked B and F. A current was sent down the wire W. This caused cathode rays to travel from C to A and hence generated X-rays at A, some of which travelled to the saucer-shaped metal plate B. The electrode F, a small Faraday cylinder, was placed at the focus of B and connected to an electrometer. The idea was that the X-rays impinging on B would release electrons and these would be gathered at F and registered on the electroscope. For this process to work properly an important condition had to be satisfied: B had to be negatively charged. Marx therefore arranged that the electric pulse which worked the X-ray tube would, by induction at D, create a small pulse in another wire which would conduct it to B. By adjusting the distances between his two vacuum tubes, and the length of the wire DB, Marx could ensure that his X-rays arrived at B at the same time as the derived impulse. The liberated electrons would then be guided to the cylinder F and the electrometer E. The maximum effect on the electrometer would be the indication that the arrival of the electrical signal and the X-rays were properly coinciding with one another. Marx then found that if he increased the distance between his two tubes — that is if he increased the length of the journey made by the X-rays — then he had to lengthen the wire DB to keep the arrival of the electrical charge on B in step with the arrival of the X-rays. A 10cm increase in the X-ray journey, say, required a 10cm increase in the wire DB. On this basis Marx concluded that the X-rays travelled with the same speed as does an electrical signal down a wire. Since he accepted that the latter speed was equal to the speed of light, he concluded that X-rays travel with the speed of light.

Figure 9.1: Marx's Apparatus for Measuring the Speed of X-rays

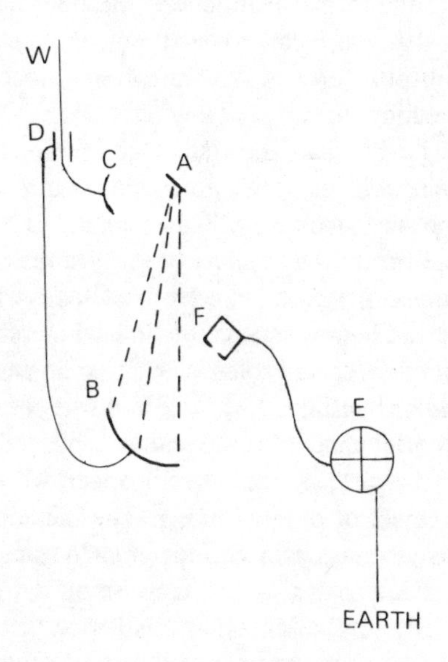

Source: *Philosophical Magazine*, vol. 14 (1907), p.447.

The important thing to notice about Marx's clever piece of apparatus is that it does not, by itself, determine the conclusion that Marx drew from it. The conclusion depends on many assumptions. In effect, the conclusion only follows if one accepts a certain story about what is actually happening in the apparatus. Wheaton calls attention to the fact that one vital step in the reasoning was the idea that the relevant electric effect was propagated down the wire DB with the speed of light. But the reasoning also depended on another assumption: that what is being picked up at B really is the effect of X-rays, rather than the effect of something else which accompanies the rays or is mixed in with them. In other words, the reasoning proceeds as it does because of the assumption that the effects being studied are unitary or homogeneous or simple. If this were to be questioned or abandoned, the meaning of the experiment would change. We shall see later that W.H. Bragg [2] (1862–1942) questioned precisely this point. Bragg thought that X-rays were particles that moved much more slowly than light. He therefore argued that Marx was picking up a side-effect. He said that when real X-rays impinged on B they would produce effects that would not behave in the way Marx assumed. Instead of the slowly moving electrons that might indeed be registered on E, they would produce highly energetic, very fast electrons that would be scattered in all directions, despite the charge on B and its special shape that was meant to encourage movement to F. The low-energy, well controlled effects that were being picked up must be due, he said, to something else.

It is interesting to compare the kind of assumption that Bragg detected in Marx's reasoning with those that are found in our common-sense thinking outside the physics laboratory. Reasoning in everyday life frequently depends on simplifications and stereotypes. We reason in terms of pre-processed categories. We say to ourselves that this is what can be expected of, say, doctors, or shop-assistants, or 'left-wingers'. Sometimes these typifications refer to accepted roles, as with 'shop-assistant'; sometimes they are typifications sustained by different sorts of interests, as when a newspaper editorial condemns the real or supposed behaviour of left-wingers. Clearly, something similar can happen when an experiment in physics is the subject of our reasoning. The meaning and plausibility of the conclusion depends on a taken-for-granted category being imposed on the process under analysis. Just as those on the receiving end of a stereotype in daily life often smart under its crudity and lack of discrimination, so we may suspect that all sorts of subtle differences might be overlooked by our scientific categories — differences that may only come to light when a critic like Bragg wants to make out a case for a rival interpretation. Until then our categories will seem to fit or describe reality in a comfortable way.

This comparison between scientific categories and those of daily life is not meant as a criticism of either. What is being pointed out is not an intrinsic defect in thinking, so much as an intrinsic necessity. We could go on drawing distinctions for ever. At some stage we must stop and accept a way of classifying and grouping things. Marx stopped at one place; Bragg wanted to draw one more distinction before he stopped. The point, however, is that in all cases we can look upon the classification that is being shared or contested or proposed as something like an institution or a convention. Underdetermination is only solved by collective agreement.

The Organisation of the Scientific Profession

Now let us sharpen the focus of the discussion by moving from the currents of accepted opinion that are diffused through a group of scientists, to the way in which that group is organised. In order to discuss this subject I will change my example from the history of X-rays to the history of chemical atomism. Terry Shinn's striking account of the reception of atomic theory in France shows the importance of the structure of authority in the scientific community.[3] His starting point is the fact that while atomic theory had been widely accepted by British, German and American chemists by 1860, it was not generally accepted by their French colleagues until approaching the turn of the century. Indeed, Shinn reports that atomic theory did not find its way into school text-books in France until the eve of the First World War. Instead of relating their experimental findings to the hypothetical behaviour of atoms, French chemists rationalised their procedures by the notion of 'equivalents'. Equivalents were empirical measures of combining weights and volumes, usually related to oxygen as a standard. They provided a way to summarise observable facts. This avoided the need to refer to hypothetical entities like atoms which could not be brought under the same kind of scrutiny as could the chemical substances on the laboratory bench.

Fortunately it is not necessary to decide the rational merits of the strategy of caution adopted by the French. We need only appreciate two facts. First, principled arguments could be made out in favour of their position. After all, French chemists had long accepted the discipline of this 'positivist' methodology and had made important contributions using the idea of equivalents. Second, we need to accept that, on the other side, most chemists had decided that the dangers of constructing a speculative atomic theory were offset by its gains. It had enabled them to clarify obscurities in their nomenclature, to construct new and useful classificatory devices, and to perceive subtle analogies between the behaviour of different chemicals.

So the question remains: why were the French out of step? Shinn answers this question by showing how French chemistry was so organised that a small number of individual scientists could exercise enormous power over the profession. These individuals had all done their most creative work using equivalents rather than atoms and, with very few exceptions, saw no reason to admit new methods into the fields upon which their reputations had been built. As a measure of the strength of this opposition Shinn cites the case of Henri Saint-Clair Deville (1818–81), a leading chemist, who forbade all discussion of atoms in his courses at the Ecole Normale Supérieur throughout the 1850s and '60s. Men like Deville had the power to impose their preferences because of the centralised control to which French education was subject and because of a remarkable practice known as 'cumul'. Cumul was the system whereby a prominent scientist would hold a whole string of influential positions. Among others, Shinn cites J.B.A. Dumas (1800–84) as an example. Dumas held the chair of chemistry at the Ecole Polytechnique, *and* at the Sorbonne, *and* in the Faculty of Medicine in Paris. Added to this he was the president of a prestigious scientific society and a chairman of an important scientific commission. Nor did cumul end with academic appointments: it spilled over into government positions. Thus Dumas was elected to the French Legislative Assembly and became minister for commerce and agriculture. He was president of the municipal council in Paris and became a senator and director of the mint. He also sat on the Conseil Supérieur de l'Instruction Publique and, for a while, acted as its president.

Dumas may have been an extreme case, but he was not unique. Shinn identifies a group of some nine chemists who, through the practice of cumul and because of the power that the bureaucratic structure placed in their hands, effectively controlled the discipline. Their ability to determine the contents of courses and text-books and access to laboratory facilities allowed them to exclude both people and ideas with which they did not agree. Shinn describes the personal ostracism and obstruction confronting those like C.F. Gerhardt (1816–56) and A. Laurent (1807–53), who directed their work away from the system of equivalents and towards atomic theory.

Examples of this kind must be interpreted with some caution. It is tempting to leave behind the impartial ground of description and adopt an evaluative stance. This is, in fact, what Shinn does — though he is very careful to separate his evaluations from his descriptions. What we have in these examples, he says, is a case where 'extra-scientific beliefs and arrangements' impinged on science and introduced 'scientifically irrelevant criteria into scientific institutions' (p.554). Although the argument is not developed in detail, the picture is of political and social factors intruding into science, distorting the scientists' grasp of the world and impeding their progress. The moral that Shinn draws is that science is best done in a plurality of scientific institutions that are 'administratively and politically . . . independent of each other'. It is easy to sympathise with this response, and it may be the right one. But before we could be confident in such an evaluation, it would be necessary to ensure that the positive or negative influence of the social variables in question had been subject to a properly controlled study. This is a remarkably difficult condition to satisfy. The information that we need, but do not possess, is how often the forms of scientific organisation that we favour have helped to sustain developments of the kind we admire, and how often forms of organisation of which we disapprove really have sustained error or retarded progress. Just think how much more difficult it would have been for Shinn to make his case if the contingencies had been different. Suppose the French had been Atomists. Their rigid bureaucracy would then have provided them with a bulwark for defending this truth against the forces in science that would have dissipated their hard-won insight. (This, after all, is how things probably looked to those who were defending equivalents.) Perhaps for the moment we had better gather data rather than lay down policy about centralisation versus pluralism in science.

Interests and Exemplars

The French chemists described by Shinn were not just defending a high-level theoretical preference — a body of ideas. They were defending a particular way of practising their science. The emphasis on empirical relations of equivalence would be embodied in their habits of thought and the detailed manipulative routines of their laboratory practice. The problems of changing a body of practices must not be underestimated. Think, for example, of the problems that would be expected if a large commercial enterprise changed its office routine or its accounting system. As well as the psychological problems of mastering new skills there would be another level at which difficulties would be generated. Much of the effort that went into developing the old procedures would lose its relevance, so past achievements would be devalued. Old relations of power and authority would be changed or subject to strain. Spheres of influence, chains of command, powers of discretion and regions of autonomy would have to be constructed anew.

One way of summarising these facts is to say that any set of practices — including scientific ones — become the focus of an interest in their preservation. For example, the investment of time and energy in a set of scientific procedures — it might be the use of a particular piece of apparatus — will generate an interest in their continuation. In the case of the French chemists these interests were massively reinforced by certain remarkable institutional arrangements. There is, however, no reason to assume that such interests will only arise in these circumstances or that they require such forces to sustain them. Interests of some kind would seem to be pervasive. The mechanism of their production is very simple and would be expected to operate even where science is organised according to open and pluralistic principles. Here, then, is a way to understanding the fine structure of the scientific enterprise whatever its large-scale organisation.

Before turning to examples it is worth noticing that vested interests in a body of scientific achievement do not operate merely in a negative way. As well as helping to illuminate resistance to change, they can also help to show why choices are made and why ideas develop in one direction rather than another. The general phenomenon of underdetermination warns us that there will typically be more than one way to understand a novel observation. If one of these ways better enables scientists to integrate the new result into existing practices, or allows scientists to maximise the use of familiar techniques and models, then here is a plausible explanation of why it will be preferred to its rivals. Innovations of this kind can serve interests just as effectively as the strategy of unyielding resistance described by Shinn. Approaching science by using sociological categories such as 'interest' clearly reinforces the idea that scientific problems are typically addressed by relating new findings to existing problem solutions, or what Kuhn calls 'exemplars'.[4]

Particles and Waves as Institutions

To provide ourselves with examples of scientific growth that are explicable in the way just described, let us go back to W.C. Röntgen's (1845–1923) X-rays. When confronted by these novel emanations physicists faced the task of relating them to existing bodies of knowledge. The problem was to find analogies that would provide channels through which expertise could flow. The available resources that held out most promise, on this as on so many occasions, were the highly elaborated techniques for understanding, respectively, the behaviour of particles and waves. Crudely, the choice was between assimilating X-rays to matter or to light.

By 1906 the standard approach was to see X-rays as an electromagnetic impulse. Clearly X-rays were rather like light — X-ray photographs seemed to clinch this. But they were not quite like light, showing negligible diffraction, refraction and reflection. While it was tempting to think of them as electromagnetic waves of extremely short wavelength, the circumstances in which X-rays were produced gave the impulse theory its initial plausibility. The new rays were produced by the impact of cathode rays, and these could usefully be thought of as charged particles. The impact of one of these particles might be expected to produce a single dislocation in the aether that spread out from the point of impact — in short, a pulse. Virtuoso exponents of electromagnetic theory, such as

J.J. Thomson (1856–1940), began to articulate and modify the wave theory in order to clarify the nature of these impulses. Wheaton describes the subtle modifications that had to be introduced into classical electromagnetic theory because the usual definitions of wavelength, frequency and energy could not be routinely applied to pulses because of their aperiodic nature. (The crucial variables came to be the width and number of the pulses.) These expedient shifts were introduced without an undue feeling of strain. Much of the confidence in the impulse approach derived from the fact that C.G. Barkla (1877–1944), a former pupil of Thomson's and a fine experimenter, was able to demonstrate that X-rays could be polarised — a classic wave-like effect.

Wheaton draws attention to the interesting fact that the result of Barkla's success was not a gain in credibility for all or any wave theory of X-rays, but only for the impulse theory. Inductive confidence was channelled and structured, not by the abstract possibilities inherent in the situation, but by the particular opinion that prevailed locally. Mediated by the impulse theory, the analogy between X-rays and light was clearly proving fruitful.

Not everyone, however, felt that the analogy with light was as appropriate or as successful as it seemed. The compelling power of the evidence for the impulse theory would seem to have diminished roughly in proportion to the distance from its source in Cambridge. Wheaton thinks it was no coincidence that the main opposition should come from W.H. Bragg, who had been a professor in distant Adelaide for some twenty years. As someone who had done little original research during that time, Bragg was perhaps somewhat peripheral to the community of physical scientists. This geographical and social distance, suggests Wheaton, was what permitted Bragg to formulate a bold alternative to the impulse approach. Bragg's return to active research was in the field of radioactivity, studying the range of penetrating power of alpha and beta rays, and their capacity to ionise gases. Because alpha and beta rays could be deflected by magnetic and electric fields, it was natural to think of them as charged particles. One of the main questions in this area concerned the nature of the third component of radioactivity, the gamma ray. In some respects they were clearly like X-rays. But rather than analyse them in terms of the impulse theory, Bragg naturally capitalised on his experience with alpha and beta particles and saw the behaviour of gamma-rays as particle-like. Past successes with an approach naturally creates a presumption in its favour. Bragg argued that because gamma rays were always produced in conjunction with alpha and beta rays, the greater penetrating power of the gamma rays, and their lack of response to deflecting fields could be explained by supposing them to be made up of alpha and beta particles combined in a neutral pair. The known analogy of X-rays and gamma-rays then became a reason for wondering if X-rays were also streams of particles rather than transverse, pulse-like vibrations. Because he thought of radioactivity in terms of particles, Bragg mentally drew a boundary between alpha, beta and gamma radiation on the one hand, and light on the other. Light, he thought, certainly *was* a wave; but X-rays were on the particle side of the boundary.

This example, even when sketched in such simplified terms, seems to bear out the suggestions made above about the importance of exemplars and the vested interests that grow up around them. A background of research in a given area does not leave scientists indifferent in their responses to novelty. They will deploy their available expertise, and this will naturally incline them to assess analogies and similarities somewhat differently from those with different backgrounds and commitments. They will see the promise inherent in two approaches differently and diverge in their judgement of how serious the known difficulties are going to prove. To those who approached X-rays from the direction of electromagnetic waves, Barkla's polarisation finding was compelling evidence in its favour. For Bragg this 'so-called polarisation' could be accounted for (in principle) by supposing that his neutral pair of particles could be made to rotate in a plane and that their absorption in, or ejection from, an atom depended on this alignment. When Barkla discovered that substances bombarded by X-rays gave out a characteristic form of secondary X radiation, this spelled more trouble for Bragg. The result made sense on the impulse theory. The pulse set the whole atom in vibration, and the electrons in the atom would resonate at a frequency determined by their internal arrangement in the atom. These oscillating charges would be the source of the characteristic X-ray spectrum of the material. For Bragg the problem was that the scattered beam of neutral pair particles should not really depend at all on the nature of the material responsible for the scattering.

If Bragg's approach had its difficulties, it also had its strengths — at precisely the point where the impulse theory was most vulnerable. A pulse spread like a wave. So when X-rays passed through a gas, the wavefront would sweep across it and spread its influence over large numbers of molecules. How was it, then, that so few molecules were ionised? And if the energy was spread across a wavefront, why was it not dissipated rather than, as seemed to be the case, being available in a surprisingly concentrated form? Didn't this all point to a mechanism more like particle collision? On these and other issues, both sides minimised their own problems and maximised those of their opponents. Their judgements concerning the weight and importance of pieces of evidence were systematically different. Those on each side of the controversy betrayed a systematic structure to their judgements not unlike those at work in the French chemists who opposed atomic theory, though, of course, in a more moderate form. In the case of the Barkla-Bragg controversy, however, there would seem to be no question of pressures or institutions outside science distorting the judgements of the men concerned. What pressures there were, were generated entirely within the structure of the profession itself and were created by such internal facts as the distribution of expertise, and the history of successful involvement with particular sub-fields and their associated theories — although the proponents in this controversy were hardly disinterested, nor were their opinions distorted.

In the light of these considerations it is worth asking how the dispute over the wave or particle nature of X-rays ended — in as far as it has ended. Although Bragg was robust in the way he pressed the case for a particle theory, he was also pragmatic. Defending the heuristic power of the particle model, he acknowledged from the outset that its scope as an account of the nature of radiation was bound to be limited. Light needed a wave model. He also felt bound to concede that, even in the area of X-ray work, it was possible that an impulse model might have some limited application. Here Bragg cited Marx's experiment indicating that X-rays travelled with the speed of light. His response was to say that perhaps some pulses were mixed in with the particles. (Notice again that Marx's experiment — which might have supported either an impulse or a full wave theory — was read as confirming the impulse theory.) This willingness on Bragg's part to share out territory seemed to increase when he moved back from Australia and took up a chair in England. When, in 1912, reports came through that W. Friedrich (1883–1968) and P. Knipping (1883–1935) had produced interference effects by passing X-rays through crystals, the situation was further altered in a dramatic way. The boundary between X-rays and light, which Bragg had been defending and exploiting, seemed to crumble. Following Einstein (1879–1955) this could have been read as a basis for saying that light, too, was made of particles. Bragg, along with most physicists, inclined to the opposite conclusion: that X-rays were, after all, best thought of as waves. Bragg and his son rapidly exploited the new techique of passing X-rays through crystals and became leaders in the field. Perhaps it was the existence of this attractive and promising line of experimental enquiry — the existence of a technique ready to be exploited — that caused the X-rays/light analogy to be read in the way it was. Profoundly difficult theoretical problems confronted those who would see the similarity as a case for saying that light was particulate; a wealth of empirical results beckoned those who had the ingenuity to develop the experiments on X-ray diffraction and to treat X-rays as fully wave-like. Even so, Bragg did not abandon the conviction that *some* features of X-ray behaviour, such as ionisation, still indicated that there was scope for a particle model.

Concluding Remarks

It is not possible within one chapter to convey the range of material in the history of science that is amenable to sociological analysis, or the scope of the work that has been done. Whole areas must be passed by with no more than a comment. For example, there are many studies concerning what is called the 'social use of nature'. These have shed light on some long-standing disputes in the history of matter theory concerning the relation of force and matter.[5] Other work has immediate relevance to the scientific theories that are being debated at this very moment. For instance, the idea used in the analysis of the Barkla-Bragg controversy — that of professional vested interests — has been used to great effect to illuminate the triumph of 'charm' over 'colour' in elementary particle physics.[6] Again, we could have followed the Barka story. After his important discovery of the characteristic series of the X-ray spectrum he continued with his work, and his preferred experimental techniques, to press the case for an ill-fated J-series, and then for an obscure experimental effect called the J-phenomenon. This work was confidently cast aside by most of his fellow scientists, and the story is a revealing example of how deviance is treated by the scientific profession.[7]

There are also some fascinating but different questions concerning 'revolutions' in science. The problem is: why are some changes seen as deep shifts in underlying assumptions, while other changes pass almost unnoticed? The obvious answer is that some changes really *are* deep ones. But this cannot be the whole story — and it may not even be any part of it. The snag is that when we want to see similarities and continuities between ideas, there are usually ways to smooth over the differences and translate from one idiom to the other. Conversely, if we want to fasten on a difference we can usually find one that can be made out to be important. (Think here of the history of theology and sectarian strife.) We have seen how the impulse theory required shifts in the definitions of frequency and energy. Could this have been seen as a revolutionary change? Could the new mixture of continuous waves and discontinuous impulses have been seen as threatening the edifice of classical physics, blurring the distinction between wave and particle and straining our everyday categories of understanding to the point of collapse? Perhaps not, but we do not yet know why this is so. Nor do we fully know why this phase of the history of physics contrasts so sharply with the situation a few years later, when a sense of crisis did appear to grip certain parts of the physics community. If the basis for the scientific community's sense of continuity and discontinuity, confidence and crisis, is still unclear, there is nevertheless reason to think that a significant part of the story will involve social variables.[8]

Notes

1. B. Wheaton, *The Tiger and the Shark. Empirical Roots of Wave-Particles Dualism* (Cambridge Univeristy Press, Cambridge, 1983)
2. W.H. Bragg, 'On the properties and natures of various electrical radiations', *Philosophical Magazine*, vol. 14, 1907, pp.429-49.
3. T. Shinn, 'Orthodoxy and Innovation in Science: The Atomist Controversy in French Chemistry', *Minerva*, vol. XVIII, no. 4 (1980), pp.539-55.
4. T.S. Kuhn, *The Structure of Scientific Revolutions* (University of Chicago Press, Chicago, 1962)
5. S. Shapin, 'History of Science and its Sociological Reconstructions', *History of Science*, vol. XX (1982), pp.157-211.
6. A. Pickering, 'The Role of Interests in High-Energy Physics. The Choice between Charm and Colour', in K. Knorr, R. Krohn and R. Whitley (eds), *The Social Process of Scientific Investigation. Sociology of the Sciences Yearbook, Volume IV, 1980* (Dordrecht, Reidel, 1981), pp.107-38.
7. B. Wynee, 'C.G. Barkla and the J Phenomenon — A Case Study of the Treatment of Deviance in Physics', *Social Studies of Science*, vol. VI, nos 3/4 (1976), pp.307-47.
8. P. Forman, 'Weimar Culture, Causality, and Quantum Theory, 1918-1927: adaptation by German Physicists and Mathematicians to a Hostile Intellectual Environment', *Historical Studies in the Physical Sciences*, vol. 3 (1971), pp.1-115.

CONTRIBUTORS

David Bloor is Reader in the Science Studies Unit of the University of Edinburgh. In 1976 he published *Knowledge and Social Imagery* and in 1983 *Wittgenstein: A Social Theory of Knowledge*.

Rom Harré is a Fellow of Linacre College, Oxford, and University Lecturer in the Philosophy of Science. He is the author of *The Principle of Scientific Thinking, Causal Powers* (with E.H. Madden), *Great Scientific Experiments* etc. and editor of *Scientific Thought 1900–1960.*

Edward Hussey is a Fellow of All Souls College, Oxford, and is University Lecturer in Ancient Philosophy at Oxford University. He has published *The Presocratics* (1972) and *Aristotle's Physics, Books III and IV* (1982).

David Knight has taught the History of Science at the University of Durham since 1964. He is Editor of the *British Journal for the History of Science;* his books include *Atoms and Elements*, (1970), *The Nature of Science* (1976), and *Ordering the World* (1981).

Pio Rattansi is Professor of the History and Philosophy of Science at University College, London. He has written extensively on science and medicine in the seventeenth century. His publications include *Isaac Newton and Gravity* (1974)

John Roche is a member of Linacre College, Oxford. He teaches History of Science and Physics. His major research interest lies in the application of the history of science to the clarification of the concepts of modern physics.

INDEX